Applied Electron Microscopy
Angewandte Elektronenmikroskopie

Band 7

Applied Electron Microscopy
Angewandte Elektronenmikroskopie

Band 7

Prof. Dr. Josef Zweck (Hrsg.)

Christian Hurm

Towards an unambiguous Electron Magnetic Chiral Dichroism (EMCD) measurement in a Transmission Electron Microscope (TEM)

Logos Verlag Berlin

Applied Electron Microscopy
Angewandte Elektronenmikroskopie

Herausgeber:
Prof. Dr. Josef Zweck
Institut für Experimentelle und Angewandte Physik
Universität Regensburg
93040 Regensburg
Germany
Email: josef.zweck@physik.uni-regensburg.de

Bibliografische Information der Deutschen Nationalbibliothek

Die Deutsche Nationalbibliothek verzeichnet diese Publikation in der Deutschen Nationalbibliografie; detaillierte bibliografische Daten sind im Internet über http://dnb.d-nb.de abrufbar.

ISBN 978-3-8325-2108-0
ISSN 1860-0034

Logos Verlag Berlin GmbH
Comeniushof, Gubener Str. 47,
10243 Berlin

Tel.: +49 (0)30 / 42 85 10 90
Fax: +49 (0)30 / 42 85 10 92
http://www.logos-verlag.de

Towards an unambiguous Electron Magnetic Chiral Dichroism (EMCD) measurement in a Transmission Electron Microscope (TEM)

Dissertation zur Erlangung des Doktorgrades der Naturwissenschaften (Dr. rer. nat.) der Fakultät Physik der Universität Regensburg

vorgelegt von

Christian Hurm

aus Regensburg

durchgeführt am Institut für
Experimentelle und Angewandte Physik
der Universität Regensburg
unter Anleitung von
Prof. Dr. J. Zweck

Oktober 2008

Promotionsgesuch eingereicht am: 04.09.2008

Tag der mündlichen Prüfung: 11.12.2008

Die Arbeit wurde angeleitet von Prof. Dr. Josef Zweck

Prüfungsausschuss:
Prof. Dr. Gunnar Bali (Vorsitzender)
Prof. Dr. Josef Zweck (1. Gutachter)
Prof. Dr. Werner Wegscheider (2. Gutachter)
Prof. Dr. Franz J. Gießibl (Prüfer)

Eine wirklich gute Idee
erkennt man daran,
dass ihre Verwirklichung von vorne herein
ausgeschlossen erscheint.

Albert Einstein, * 14. März 1879, † 18. April 1955

Contents

1. Introduction - the ChiralTEM project

In mineralogy, *dichroism* is known since several hundred years. In the original meaning, dichroic (deduced from the Greek *dikhroos*, two-colored) refers to the optical characteristic of an item that splits a beam of light in two beams with different wavelengths. An impressive example for such a material is Tansanit ($Ca_2Al_3(SiO_4)_3OH$), as can be seen in figure 1.1.

A more particular meaning of *dichroic* refers to the attribute of some materials that have a different ability of absorbing light, depending on the polarization of this incident light. This definition of *dichroic* is independent of the original meaning of the word (two-colored) and was first described by Peter Debye in 1939. Even more exceeding is the so called magnetic dichroism - the degree of absorption of polarized light depends on the direction of the magnetization in a specimen. For circular polarized light, this effect was first predicted in 1975 by Erskine and Stern [Ers75]. After the first experimental verification of this effect, using circular polarized X-rays in 1987 by Schuetz [Schu87], this technique became a popular tool for micromagnetic investigations, called XMCD[1] (the theory of XMCD is discussed more detailed in chapter 2). A description of XMCD by sum rules was given by Thole and Carra in 1992/93 ([Tho92], [Car93])

After the statement of the wave-particle dualism by De Broglie in 1924 [Bro24], many different attributes of (optical) light were found also for electron waves, for example in 1955 by Moellenstedt and Dueker the possibility of interference [Moe56]. In contrast to these analogies, it was commonly thought until the beginning of the 21th century that magnetic circular dichroism was not possible for electrons. But Peter Schattschneider postulated this effect in 2003, exclusively based upon formal analogy, and named it EMCD[2] in analogy to XMCD [Scha03]. On the basis of this proposal, an international research project under the lead of Peter Schattschneider (Vienna), called ChiralTEM, was brought into life. Major task of this project was the experimental verification of the predicted effect. The participating bodies can be seen in figure 1.2.

Within the ChiralTEM project, the group of Josef Zweck (Regensburg) had three responsibilities:

1. *Preparation and characterization of suitable specimens*

2. *Manipulation of the specimen's magnetization*

[1]X-ray Magnetic Circular Dichroism
[2]Electron (Energy Loss) Magnetic Chiral Dichroism

Fig. 1.1.: *Example for a dichroic mineral: Tansanit (same crystal, regarded under two different angles) appears in two different colors, depending on the viewing angle [Lav08].*

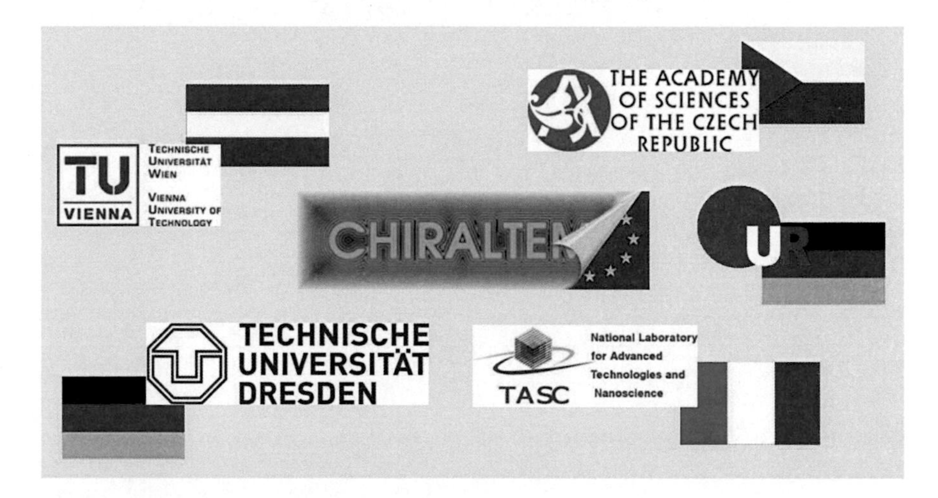

Fig. 1.2.: *Participants of the ChiralTEM project:*
Technische Universitaet Wien, Prof. Schattschneider
Technische Universitaet Dresden, Prof. Lichte
Trieste Laboratory for Advanced Technologies and Nanoscience, Prof. Carlino
University of Regensburg, Prof. Zweck
Academy of Sciences, Prague, Prof. Novak
The project was financed by the European Union under the contract number 508971.

3. *Dissemination of knowledge*

The last responsibility, the dissemination of knowledge, was covered for example by the creation and maintenance of the project homepage [Chi08], the organization of a scientific workshop within the 6th Microscopy Conference in Davos (Switzerland) in 2005 and the carrying out of several public relations, for example an informal meeting at the 71st Annual Meeting of the DPG[3] in Regensburg in 2007 and are not explicitly discussed in this work.

The responsibilities of the other groups were (according to the contract of the EU, contract number 508971)

- **Vienna** - project management, experiments, specimen preparation, theoretic framework
- **Dresden** - beam manipulation using a electrostatic bisprism
- **Prag** - theoretic framework, calculations on EMCD
- **Trieste** - comparison EMCD - XMCD

Within the ChiralTEM project, the magnetic dichroism for electrons was confirmed [Scha06]. A timetable comparison between XMCD and EMCD is given in table 1.1. Of course, this timetable appears somewhat unfair as plenty of analogies between the two techniques were used. But nevertheless it emphasises quite impressive the achievements of the ChiralTEM project.

	XMCD		EMCD	
Theory / Prediction	1975	Erskine, Stern	2003	Schattschneider
first experiment	1987	Schuetz	2003	Rubino
Sum rules	1992/1993	Thole, Carra	2007	Rusz, Calmes

Table 1.1.: *Comparison of the time from the first prediction over the experiment to the theoretic description of XMCD and EMCD [Rub07a].*

At the beginning of this project, it has not been scheduled to perform dichroic experiments in Regensburg. Unexpectedly, the TEM in Regensburg has been equipped with an energy filter in 2007. Thus it was fortunately possible to measure the dichroic effect also in Regensburg within the framework of the present dissertation. In the present work, the results of the project tasks *Preparation and characterization of suitable specimens* and *Manipulation of the specimen's magnetization* are published.

In the first part (chapters 2 and 3), the theory of EMCD and the required experimental setup is curtly outlined. The second part (chapters 4 and 5) evaluates different specimen

[3]Deutsche Physikalische Gesellschaft

preparation techniques and the magnetic properties of different materials concerning the suitability for dichroic experiments. These experiments, on the one hand the verification of the dichroic effect in a TEM and on the other hand the verification of its magnetic origin by an in-situ reversal of the specimen's magnetization are described in the third part (chapters 6 and 7). Finally, a perspective of possible future developments and a summary of this work is given in the fourth and last part (chapters 8 and 9).

2. Theoretical considerations on Electron Magnetic Chiral Dichroism (EMCD)

The theory of Electron (Energy Loss) Magnetic Chiral Dichroism (EMCD) was first described in 2003 by Peter Schattschneider [Scha03]. Although the focal point of this dissertation is not the theoretic explanation of this effect, but the facilitation of the experimental verification, some fundamentals are necessary for a conclusive argumentation. These fundamentals are given in this chapter, starting with a short explanation of XMCD. After a discussion on the formation of electron energy loss spectra, the formal analogy between EMCD and XMCD is demonstrated. Therefrom, the requirements for EMCD measurements are defined, reverting to a small excursion to the dynamic diffraction theory. The conclusion of the present chapter will be a facile perspective on the possibility of a theoretic prediction of the dichroic signal - showing only the results that are important for the following chapters.

2.1. From XMCD to EMCD

2.1.1. Functionality of XMCD

The first experimental confirmation of XMCD was done in 1987 [Schu87] on the iron K-edge. The information of XMCD is based on an absorption coefficient for the probing X-rays, depending on the helicity of their circular polarization. Generally, the superposition of two linear polarized waves, with a phase shift $\Delta\phi \neq 0$ in between, creates an elliptical polarized wave. In the special case of a phase shift between the two partial waves of $\Delta\phi = \frac{\pi}{2}$ (and identical amplitudes), the vector of the electric field is rotating at right angle around the propagation vector with a constant absolute value. This case is called circular polarized light [Nie93] (compare figure 2.1), and the polarization is called *right circular polarized* or *RCP* if the helicity follows a right hand rule, otherwise it is called *left circular polarized* or *LCP*.

In this work, the most interesting specimens will be iron, cobalt or nickel films - altogether $3d$ transition metals with a strong magnetic circular dichroism at the L resonances ($2p \rightarrow 3d$). To calculate the absolute value of the dichroic effect, one has to consider all possible transitions from the $2p$ levels to the $3d$ continuum that are allowed according to the selection rules, especially the rule $\Delta m = \pm 1$. In this context, transitions with $\Delta m = +1$ are only possible for RCP light and otherwise LCP light provides only transitions with $\Delta m = -1$.

To calculate the dichroic difference between two spectra at a certain edge, one has to consider the occupation probabilities of the initial states and the transition probabilities

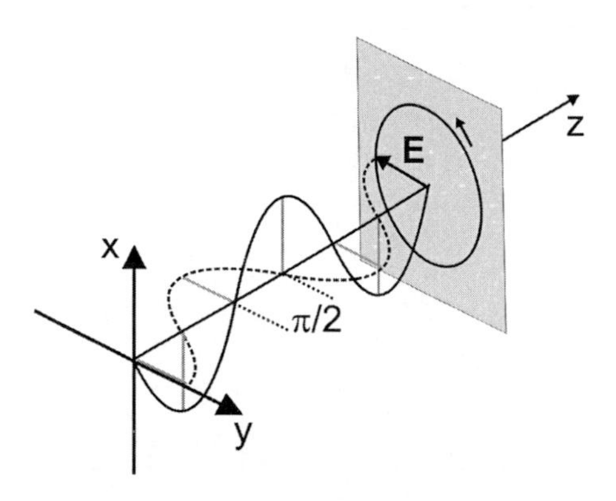

Fig. 2.1.: *The superposition of two linear polarized waves with a phase shift of $\frac{\pi}{2}$ in between results in a circular polarized wave (cognizable at the rotating vector of the electric field $\vec{E}$).*

for all the allowed transitions separately for RCP and LCP light. Thereby, the transition probability $p_{f,i}$ from an initial state $|i\rangle$ (with an energy E_i) to a final state $\langle f|$ (with an energy E_f) is given by Fermi's golden rule by [Huf03]

$$p_{f,i} = \frac{2\pi}{\hbar} \left|\langle f|H|i\rangle\right|^2 \delta(E_f - E_i - \hbar\omega) \tag{2.1}$$

with a photon energy $\hbar\omega$. As apparent, the delta function allows only contributions if the difference between E_f and E_i is exactly the photon energy (at least for T=0). The Hamiltonian H for the interaction of the incident photon with an electron is given as

$$H = \frac{e}{mc} \vec{A} \cdot \vec{p} \tag{2.2}$$

with the electron charge e, the electron mass m and the speed of light c. $\vec{A}$ is the vector potential of the electromagnetic radiation and $\vec{p}$ is the momentum operator. For a paramagnetic specimen, the transition probability is symmetric for RCP and LCP light [Hop05]. In a ferromagnet, the energy is reduced for electrons with a magnetic spin moment oriented parallel to the exchange field and thus the d band is split into a spin-up band and a spin-down band. The density of states (DoS) of the spin-down states is shifted relative to the spin-up states in the direction of the energy axis. The two shifted partial bands have different occupation and (in general) also a different density of free (unoccupied) states above the Fermi energy. Thus, the excitation probability is not symmetric for left circular polarized and right circular polarized photons due to a different density of free states above the Fermi energy. This is sketched in figure 2.2.

The largest dichroic signal is achieved if the polarization vector $\vec{\sigma}$ is parallel (or antiparallel) to the specimen's (local) magnetization $\vec{M}$. If this condition is not fulfilled, the

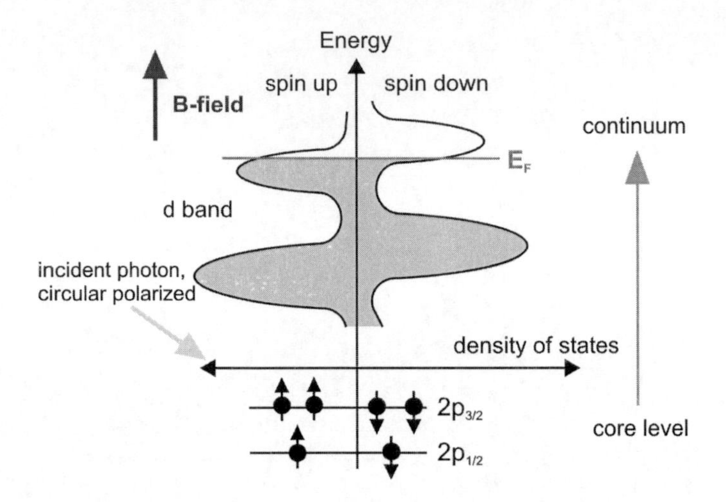

Fig. 2.2.: *The absorption of a photon excites a core electron from a 2p state to a d state in the continuum above the Fermi energy E_F. The excitation from the $2p_{1/2}$ level is called L_2-edge and the excitation from the $2p_{3/2}$ level is called L_3-edge. A magnetic field in "spin-up" direction shifts down the density of states (DoS) of the spin-down states relative to the spin-up states. Thus, the excitation probability is not symmetric for left circular polarized (LCP) and right circular polarized (RCP) photons due to a different densities of free states above the Fermi energy.*

(reduced) intensity I_{XMCD} of the dichroic signal can be estimated according to [Scho02] to be

$$I_{\mathrm{XMCD}} \propto \left|\vec{M}\right| \cdot \cos\phi(\vec{\sigma}, \vec{M}), \tag{2.3}$$

with ϕ as angle between the two vectors, $\vec{\sigma}$ and $\vec{M}$. One can see that the sign of the dichroic signal changes if either the vector of the circular polarization or the magnetization of the specimen changes sign. Assuming a similar behavior for EMCD, the magnetization of the specimen has to be parallel (or antiparallel) to the electron's direction of propagation to get a maximum dichroic effect. More detailed information on XMCD can be found in [Goe01], [Hop05] or [Huf03].

2.1.2. Electron energy-loss spectra

A small insertion about the mechanism how the electrons lose their energy in a crystal specimen should lead from the previous basics of XMCD to the following formal analogies between XMCD and EMCD. The interaction between a (fast) probe electron and the specimen can be either elastic (without detectable energy loss) or inelastic (with appreciable energy loss)[Ege96]. Elastic scattering in the meaning of an electron energy loss smaller than the energy loss resolution of the spectrometer occurs mainly by a Coulomb interaction of the fast electron with the atomic nucleus (i.e. Bragg scattered electrons). The majority of the incident electrons remain unscattered or they are scattered elastically by a small angle (some 10 mrad). In an energy loss spectrum, they form the *zero-loss peak*. There are several possible reasons of *inelastic* scattering processes and only the most important ones are mentioned in the following lines.

The atoms in a crystal vibrate around their lattice positions due to their thermal energy and the zero point energy. These vibrations can act as scattering source for the probe electrons - the electrons generate and absorb *phonons*. As the phonon energy can not exceed the energy $k_B T_D$ with the Boltzmann constant $k_B = 1.38 \cdot 10^{-23}\,\frac{\mathrm{J}}{\mathrm{K}}$ and the Debye temperature T_D[1], the energy loss (or gain) due to one single phonon excitation is smaller than 0.1 eV and only effects a broadening of the zero-loss peak.

More energy is transfered by an excitation of the free electron gas. This scattering process is called plasmon[2] excitation. The energy loss of the plasmon-peak is in the range of 5 eV to 50 eV. At a given atomic number and defined experimental conditions, the specimen thickness can be calculated in multiples of the mean free path by a comparison of the zero-loss peak and the plasmon peak [Ege96]. This possibility will be helpful for the characterization of EMCD specimens (compare chapters 5 and 6).

For higher energy losses (for example due to the continuous *Bremsstrahlung* background), the intensity decreases following a power law, but it is superimposed by the element specific energy-loss due to inner-shell excitations at the ionization thresholds. Examples for electron energy loss inner-shell excitation edges can be found in table 2.1.

Element	atomic number	K-edge	L_1-edge	L_2-edge	L_3-edge
Fe	26	7113 eV	846 eV	721 eV	708 eV
Co	27	7709 eV	926 eV	794 eV	779 eV
Ni	28	8333 eV	1008 eV	872 eV	855 eV

Table 2.1.: *Some examples for the electron energy loss of ionization edges according to [Ege96].*

For an analysis of the ionization edges, the continuous background has to be removed. This can be done by fitting a power law correction-function in a suitable energy loss range before the ionization edge and subtracting this correction-function from the whole energy-loss spectrum. An example for a zero-loss peak with plasmon peak and the $L_{3,2}$-

[1] The Debye temperature for example of iron is $T_D(\mathrm{Fe}) = 470\,\mathrm{K}$.

[2] The plasmon is the quasiparticle resulting from the collective oscillations of the free electron gas [Rub07a].

edge of a cobalt specimen with a thickness of 0.8 multiples of the mean free electron path can be found in figure 2.3. In this figure, also the calculated background function and the pure edge signal after background subtraction is shown.

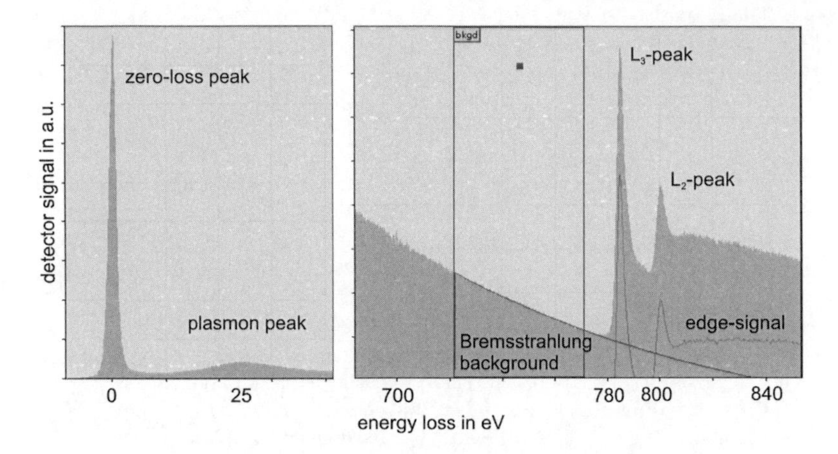

Fig. 2.3.: *Example of an energy loss spectrum of a cobalt specimen in two clippings. The left side shows the zero-loss peak with plasmon peak. The comparison of the two peaks (done by the software digital micrograph [Gat08]) delivers the specimen thickness to be 0.8 times the mean free path (MFP - the average distance between two scattering events of the same probe electron). The right side shows (in a different scale of energy loss and intensity) the L_3 and the L_2 edge with a calculation of the background function and the background corrected edge signal.*

For quantitative results it is important that the probe electron was scattered only once in the specimen - otherwise, the energy loss can not be clearly assigned to a certain scattering process. The average distance between two (inelastic) scattering events of the same probe electron is called *(inelastic) mean free path (MFP)*. The MFP is inversely proportional to a value called *(inelastic) total scattering cross section*. The scattering cross section itself contains the probability for one single atom to interact with the probe electron and will become important for the formal analogy between XMCD and EMCD in the following section 2.1.3. If the specimen is thinner than the mean free path, multiple scattering processes are rare and can be neglected in a first approximation.

2.1.3. Formal analogies

The final part of the section *"from XMCD to EMCD"* is a description of the formal analogy of the scattering cross sections of electron scattering and X-ray scattering. The ChiralTEM project was based right upon this analogy.
The absorption cross section in X-ray absorption Spectrometry (XAS) is given by ([Scha03], [Rub07a]) to be

$$\sigma = \sum_{i,f} 4\pi^2 \alpha \hbar\omega \left| \left\langle i | \vec{\epsilon} \cdot \vec{R} e^{i\vec{q}\cdot\vec{R}} | f \right\rangle \right|^2 \delta(E_f - E_i - E) \tag{2.4}$$

with fine structure constant $\alpha = \frac{1}{137}$, photon energy $\hbar\omega$, polarization vector $\vec{\epsilon}$ and $\hbar\vec{q}$ the momentum of the absorbed photon. i represents the initial state and f the final state of the transition. The δ-function only allows a non-zero absorption cross section for photons with an energy $E = E_f - E_i$, where E_f is the energy of the final state and E_i is the energy of the initial state.

In contrast to photon scattering, the double differential scattering cross section (DDSCS) for electrons can be written as

$$\frac{\partial^2 \sigma}{\partial \Omega \partial E} = \frac{4\gamma^2}{a_0^2} \frac{k_f}{k_i} \frac{S(\vec{q}, E)}{q^4} \tag{2.5}$$

with

$$S(\vec{q}, E) = \sum_{i,f} \left| \langle i | e^{i\vec{q}\cdot\vec{R}} | f \rangle \right|^2 \delta(E_f - E_i - E) \tag{2.6}$$

and the momentum transfer $\vec{q} = \vec{k}_f - \vec{k}_i$ as difference between the two wave vectors of the final and the initial state. The relativistic factor is given by $\gamma = \frac{1}{\sqrt{1-\frac{v^2}{c^2}}}$, a_0 is the Bohr radius, $\vec{R}$ is the local operator and in analogy to the photon case, E describes the energy loss. The factor $S(\vec{q}, E)$ in equation 2.5 is called *dynamic form factor (DDF)*. A detailed derivation of the double differential scattering cross section can be found in [Nel99].
The XAS notation of the absorption cross section σ (equation 2.4) defines the probability for the absorption of a photon of energy E. Except for the solid angle reference, this comprehension is equivalent to the DDSCS of electron scattering (equation 2.5) which describes the scattering probability with a certain (electron) energy loss E. The (total) cross section can be calculated from the differential cross section by integral on the whole sphere of observation ($\sigma = \int \mathrm{d}\Omega \frac{\mathrm{d}\sigma}{\mathrm{d}\Omega}$).

Using dipole approximation[3], the first term vanishes in equation 2.6 due to the orthogonality of initial and final state [Rub07a]. Plugging in the DFF, equation 2.5 becomes

[3] $e^{i\vec{q}\cdot\vec{R}} = 1 + i\vec{q}\cdot\vec{R} + O(\vec{q}\cdot\vec{R})^2$, with a small momentum transfer $\vec{q}$.

$$\frac{\partial^2\sigma}{\partial\Omega\partial E} = \sum_{i,f} \frac{4\gamma^2}{a_0^2 q^4} \frac{k_f}{k_i} \left|\langle f|\vec{q}\cdot\vec{R}|i\rangle\right|^2 \delta(E_f - E_i - E) \tag{2.7}$$

Applying dipole approximation to equation 2.4 like in the photon case, the second term of the expansion can be neglected in comparison to the first term of the series expansion. Thus, equation 2.4 can be simplified to

$$\sigma = \sum_{i,f} 4\pi^2\alpha\hbar\omega \left|\langle f|\vec{\epsilon}\cdot\vec{R}|i\rangle\right|^2 \delta(E_f - E_i - E). \tag{2.8}$$

A comparison of the two equations above, eq. 2.7 and eq. 2.8, shows a formal analogy in the transition probabilities for photon and electron scattering (compare the red marked terms in the two equations). In this notation, the polarization vector $\vec{\epsilon}$ in photon scattering directly corresponds to the momentum transfer $\hbar\vec{q}$ of inelastic electron scattering. According to this analogy, a linear magnetic dichroism for electrons was found for electron scattering in analogy to XMLD (X-ray magnetic linear dichroism). See [Yua97] or [Ake03] for details.

The proposal for the ChiralTEM project [Scha03] even takes a step forward and extends this analogy to circular dichroism. In consequent analogy to photon optics, where circular polarized light is generated by a superposition of two linear polarized waves with a phase shift of $\frac{\pi}{2}$ in between ($\vec{\epsilon}$ is replaced by $\vec{\epsilon} \pm i\vec{\epsilon}'$, remember figure 2.1), EMCD uses the superposition of two perpendicular momentum transfers. The vector $\vec{q}$ in equation 2.6 is replaced by $\vec{q} + i\vec{q}'$ with $\vec{q} \perp \vec{q}'$ and $|\vec{q}| = |\vec{q}'|$. Doing so, equation 2.7 transforms into

$$\begin{aligned}
\frac{\partial^2\sigma}{\partial\Omega\partial E} &= \frac{4\gamma^2}{a_0^2}\frac{k_f}{k_i} \sum_{i,f} \left[\frac{1}{q^4} \left|\langle f|\vec{q}\cdot\vec{R}|i\rangle\right|^2 + \frac{1}{q'^4} \left|\langle f|\vec{q}'\cdot\vec{R}|i\rangle\right|^2 \right. \\
&\quad - \frac{1}{q^2 q'^2} \langle f|\vec{q}\cdot\vec{R}|i\rangle\langle i|i\vec{q}'\cdot\vec{R}|f\rangle \\
&\quad \left. + \frac{1}{q^2 q'^2} \langle f|i\vec{q}'\cdot\vec{R}|i\rangle\langle i|\vec{q}\cdot\vec{R}|f\rangle \right] \delta(E_f - E_i - E)
\end{aligned} \tag{2.9}$$

with dynamic form factors for $\vec{q}$ and $\vec{q}'$ (first line) which describe the inelastic scattering process from two incident electron waves in two outgoing electron waves with an energy loss of E. The second and third line contains inelastic interference terms, so called *mixed dynamic form factors (MDFFs)*. These MDFFs can be written as

$$S(\vec{q},\vec{q}',E) = \sum_{i,f} \langle f|\vec{q}\cdot\vec{R}|i\rangle\langle i|\vec{q}'\cdot\vec{R}|f\rangle\delta(E_f - E_i - E) \tag{2.10}$$

$$S(\vec{q}',\vec{q},E) = \sum_{i,f} \langle f|\vec{q}'\cdot\vec{R}|i\rangle\langle i|\vec{q}\cdot\vec{R}|f\rangle\delta(E_f - E_i - E). \tag{2.11}$$

The equations 2.10 and 2.11 can be understood as interference terms - similar to the double slit experiment. Due to the interference terms, electrons can be found not only in the Bragg spots, but also in the area between the Bragg spots. For the EMCD measurement, a position in the diffraction plane with two perpendicular momentum transfers $\vec{q}$ and $\vec{q}'$ (coming from the two Bragg scattered beams as (virtual) electron sources) corresponds to the photon case ($\vec{\epsilon} \perp \vec{\epsilon}'$). In figure 2.4, this analogy between XMCD and EMCD is drawn schematically.

In the photon case, the electric field vector $\vec{E}$ is parallel to the polarization vector $\vec{\epsilon}$. Circularly polarized light has a rotating vector of the electric field that forces the electron in the specimen from ground state to an excited state, requiring a change of the magnetic quantum number m_j by $\Delta m_j = \pm 1$ depending on the helicity of the incident photon. In the electron case, the electric field generated by the passing probe electron interacts with the bulk electrons.

A comparison of the experimental setups for XMCD and EMCD is shown schematically in figure 2.4. This figure also anticipates some requirements on the experimental setup. In EMCD, the specimen has to fulfill two tasks - a coherent splitting of the electron beam into two waves with a phase shift relative to each other, and providing magnetic moments for a detection of the dichroic difference between two spectra. These requirements will be discussed in detail in section 2.2 and a suitable experimental setup is shown in section 3.3.

As the equation

$$S(\vec{q}, \vec{q}\,', E) = S^*(\vec{q}\,', \vec{q}, E) \tag{2.12}$$

is always true, the diagonal elements $S(\vec{q}, \vec{q}, E)$ (i.e. the DFFs) are real. The off-diagonal elements are complex unless the crystal has an inversion center ($S(\vec{q}, \vec{q}\,', E) = S(-\vec{q} - \vec{q}\,', E)$) **and** time inversion symmetry is not broken ($S(\vec{q}, \vec{q}\,', E) = S(-\vec{q}\,', -\vec{q}, E)$). Only in this case,

$$S(\vec{q}, \vec{q}\,', E) = S(-\vec{q}\,', -\vec{q}, E) = S(\vec{q}\,', \vec{q}, E) = S^*(\vec{q}, \vec{q}\,', E) \tag{2.13}$$

and considering all elements are real - the target system has no chirality and a (circular) dichroic effect is excluded. At the presence of (non degenerated) magnetic moments, time symmetry is broken. According to [Rub07a], equation 2.9 can be rewritten as

$$\frac{\partial^2 \sigma}{\partial \Omega \partial E} = \frac{4\gamma^2}{a_0^2} \frac{k_f}{k_i} \left[\frac{S(\vec{q}, E)}{q^4} + \frac{S(\vec{q}\,', E)}{q'^4} + 2\Im \left(\frac{S(\vec{q}, \vec{q}\,', E)}{q^2 q'^2} \right) \right] \tag{2.14}$$

and it can be shown that the chiral difference signal between two spectra, taken at different helicities of the incident electron waves, is closely related to the imaginary part of the MDFF. A more detailed reflection on this relation would go beyond the scope of the present work but the theoretic description so far is sufficient to define some requirements on an EMCD experiment based upon theory (some more practical requirements are discussed in section 5.1).

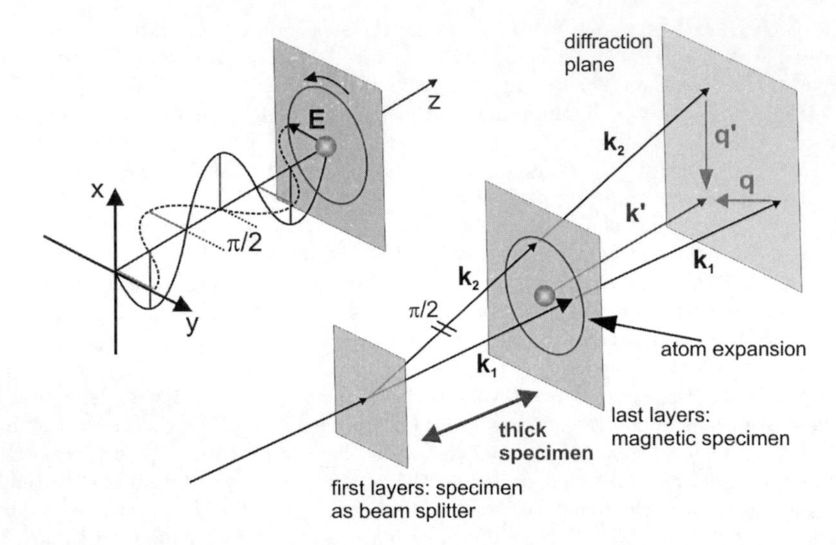

Fig. 2.4.: *In contrast to the photon case (left side), the specimen has two functions in EMCD (right side): splitting the incident electron wave (with wave vector $\vec{k}_1$) in two waves (with wave vectors $\vec{k}_1$ and $\vec{k}_2$) with a phase shift of $\frac{\pi}{2}$ in between (in the first layers of the specimen) and providing magnetic moments for a different absorption coefficient at the edges (in the last layers of the specimen). The actual measurement is done in the diffraction plane at the position of the vector $\vec{k}'$. An electron detected at this position suffered the two momentum transfers $\hbar\vec{q}$ and $\hbar\vec{q}'$.*

2.2. Requirements for EMCD experiments

So far, four requirements on an EMCD experiment can be expressed:

- According to the formal analogy, two simultaneous momentum transfers ($\vec{q}$ and $\vec{q}'$) must occur. Therefore, the incident electron wave has to be divided in two coherent partial waves in the directions of $\vec{k}_1$ and $\vec{k}_2$. In the ChiralTEM project, it was discussed to achieve this with an electrostatic biprism above the specimen plane [For05]. As this approach did not persuade due to technical limitations - up to now, the only way to split the electron wave in an applicable way, is the usage of a single crystal specimen which splits the incident wave in Bragg scattered waves. A different approach using a twin hole aperture is currently under investigation and described in section 8.2.
- The two momentum transfers $\vec{q}$ and $\vec{q}'$ must not be parallel. The largest dichroic

signal is expected for $\vec{q} \perp \vec{q}'$. This is equivalent to polarization vectors $\vec{\epsilon}$ and $\vec{\epsilon}'$ with $\vec{\epsilon} \perp \vec{\epsilon}'$ in the case of electromagnetic waves. In electron optics, this condition can be fulfilled when choosing a suitable position in the diffraction pattern as displayed in figure 2.4 or described in detail in section 3.3.

- Similar to the superposition of two linear polarized (optical) waves to a circular polarized wave (remember figure 2.1), there must be a phase shift of $\frac{\pi}{2}$ between the two partial waves in the directions of $\vec{k}_1$ and $\vec{k}_2$. Fortunately, this condition is automatically fulfilled when setting the diffration pattern to a two beam case (derived in detail in the following section 2.3).

- To receive two spectra with a dichroic difference in between, there must be the possibility to change the *helicity* of the probe electron. This can be done either by a change of the position in the diffraction plane, as described in section 3.3 and performed in chapter 6, or by a reversal of the specimen's magnetization as discussed in chapter 7.

Two examples for practicable experimental setups fulfilling all these requirements are shown in section 3.3.

2.3. Dynamical diffraction theory

According to the precedent sections, the diffraction mode of the TEM (compare section 3.1) will be a suitable tool for EMCD experiments. A more detailed insight in the interaction between the electron and the crystal can be derived by the dynamic diffraction theory, solving the Schrödinger equation

$$-\frac{\hbar^2}{2m}\nabla^2\Psi(\vec{r}) + P(\vec{r})\Psi(\vec{r}) = E\Psi(\vec{r}) \qquad (2.15)$$

for high-energy electrons with a potential energy $P(\vec{r})$ following the periodicity of the crystal lattice [Ful01]. In this context, only the idealized two beam case (in the diffraction plane, only the forward-scattered beam $\vec{\Phi}_0$ and one diffracted beam $\vec{\Phi}_{\vec{g}}$ occur[4]) will be discussed. Defining the z-direction parallel to the electron beam, these two beams can be expressed as[5]

$$\Phi_0(\vec{r}) = \frac{1}{\sqrt{V}}\phi_0(z)e^{i\vec{k}_0\cdot\vec{r}} \qquad (2.16)$$

$$\Phi_{\vec{g}}(\vec{r}) = \frac{1}{\sqrt{V}}\phi_{\vec{g}}(z)e^{i(\vec{k}_0+\vec{g})\cdot\vec{r}} \qquad (2.17)$$

[4]In practice, an ideal two beam case can not be reached because other beams will always be weakly excited, too. This restriction does not affect the statements of this section as long as the main intensity is found in the two beams Φ_0 and $\Phi_{\vec{g}}$.

[5]Within this work, 0 has also the meaning of $\vec{0}$.

normalized in the volume V of the crystal and with $\vec{k_0}$ the wave vector of the incident electron wave and $\vec{g}$ the momentum transfer of the Bragg scattered beam.

One possible approach[6] is the construction of a basis set of eigenfunctions, using Bloch waves. Bloch wave functions are steady-state solutions of the Schrödinger equation (2.15) with a translationally-periodic potential of infinite extent, well known from solid state physics [Ash01], [Kit02]. For the two beam case, a minimum of two Bloch waves $\vec{\Psi}_1(\vec{r})$ and $\vec{\Psi}_2(\vec{r})$ is required. As the beams $\vec{\Phi}_0$ and $\vec{\Phi}_{\vec{g}}$ are to arise as beats of the two Bloch waves, the wave vectors $\vec{k}^{(1)}$ and $\vec{k}^{(2)}$ of the Bloch waves have to differ slightly around a medium wave vector $\vec{k}$,

$$\vec{k}^{(1)} = \vec{k} + \gamma^{(1)}\vec{z} \tag{2.18}$$
$$\vec{k}^{(2)} = \vec{k} + \gamma^{(2)}\vec{z} \tag{2.19}$$

with (initially) different variations[7] $\gamma^{(1)}$ and $\gamma^{(2)}$, both small compared to $|\vec{k}|$. This is also known from beats of acoustic waves [Tip94]. With these wavevectors, the two Bloch waves can be written as

$$\Psi^{(1)}(\vec{r}) = \frac{1}{\sqrt{V}}\psi^{(1)}e^{i\vec{k}^{(1)}\cdot\vec{r}} \tag{2.20}$$
$$\Psi^{(2)}(\vec{r}) = \frac{1}{\sqrt{V}}\psi^{(2)}e^{i\vec{k}^{(2)}\cdot\vec{r}}. \tag{2.21}$$

As the waves 2.20 and 2.21 are assumed to be a basis set for the two beams, the latter can be expressed by a sum of these two Bloch waves[8].

$$\Phi_0(\vec{r}) = C_0^{(1)}\Psi^{(1)}(\vec{r}) + C_0^{(2)}\Psi^{(2)}(\vec{r}) \tag{2.22}$$
$$\Phi_{\vec{g}}(\vec{r}) = C_{\vec{g}}^{(1)}\Psi^{(1)}(\vec{r})e^{i\vec{g}\cdot\vec{r}} + C_{\vec{g}}^{(2)}\Psi^{(2)}(\vec{r})e^{i\vec{g}\cdot\vec{r}} \tag{2.23}$$

using constant coefficients $C_{0,\vec{g}}^{(i)}$.

Substituting the forward scattered beam (2.16), the two Bloch waves (equations 2.20 and 2.21) and the wave vectors (equations 2.18 and 2.19) into equation 2.22 and analogously the diffracted beam (equation 2.17), the Bloch waves, and the wave vectors into equation 2.23, the resulting equations are

---

[6] Other possible approaches can be found in [Ful01], [Wil96] or [Rei97], but will not lead in such a straight-forward way to the aspired results.

[7] In this notation, $\vec{z}$ is the unit vector in z-direction and $\gamma^{(1)}$ and $\gamma^{(2)}$ have the dimension of a wave vector.

[8] Unlike in equation 2.23, in equation 2.22 the exponential functions $e^{i\cdot 0\cdot\vec{r}}$ cease to a factor 1 as there is no momentum transfer for the forward scattered beam.

$$\frac{\Phi_0(z)}{\sqrt{V}} e^{i\vec{k}_0\cdot\vec{r}} = \frac{1}{\sqrt{V}}\left(C_0^{(1)}\Psi^{(1)}e^{i(\vec{k}+\gamma^{(1)}\vec{\hat{z}})\cdot\vec{r}} + C_0^{(2)}\Psi^{(2)}e^{i(\vec{k}+\gamma^{(2)}\vec{\hat{z}})\cdot\vec{r}}\right) \quad (2.24)$$

$$\frac{\Phi_{\vec{g}}(z)}{\sqrt{V}} e^{i(\vec{k}_0+\vec{g})\cdot\vec{r}} = \frac{1}{\sqrt{V}}\left(C_{\vec{g}}^{(1)}\Psi^{(1)}e^{i(\vec{k}+\vec{g}+\gamma^{(1)}\vec{\hat{z}})\cdot\vec{r}} + C_{\vec{g}}^{(2)}\Psi^{(2)}e^{i(\vec{k}+\vec{g}+\gamma^{(2)}\vec{\hat{z}})\cdot\vec{r}}\right) \quad (2.25)$$

After a division by the particular exponential functions and assuming (without loss of generality for the aspired conclusions) the symmetric case,

$$C_0^{(1)}\Psi^{(1)} = C_{\vec{g}}^{(1)}\Psi^{(1)} = C_0^{(2)}\Psi^{(2)} = -C_{\vec{g}}^{(2)}\Psi^{(2)} = \frac{1}{2} \quad (2.26)$$

the equations 2.24 and 2.25 can be transformed (see [Ful01] for more detailed calculations) to

$$\Phi_0(z) = \left(e^{i(\gamma^{(1)}+\gamma^{(2)})\frac{z}{2}}\right)\cdot\cos\frac{\gamma^{(1)}-\gamma^{(2)}}{2}z \quad (2.27)$$

$$\Phi_{\vec{g}}(z) = \left(e^{i(\gamma^{(1)}+\gamma^{(2)})\frac{z}{2}}\right)\cdot i\sin\frac{\gamma^{(1)}-\gamma^{(2)}}{2}z. \quad (2.28)$$

It can be shown that the simplification 2.26 also means $\gamma^{(1)} = -\gamma^{(2)}$. Thus, the phase factors in the equations 2.27 and 2.28 become equal 1 and the equations can be written as

$$\Phi_0(z) = \cos\frac{2\cdot\gamma^{(1)}}{2}z \quad (2.29)$$

$$\Phi_{\vec{g}}(z) = i\sin\frac{2\cdot\gamma^{(1)}}{2}z. \quad (2.30)$$

These expressions for the forward scattered beam $\vec{\Phi}_0$ and the diffracted beam $\vec{\Phi}_{\vec{g}}$ deliver two important results for EMCD experiments. First, the phase of the diffracted beam is shifted by $\frac{\pi}{2}$ in relation to the forward scattered beam as can be seen by the factor i in the diffracted beam (equation 2.30). This is one of the requirements of the conclusion by analogy (chapter 2.2), which is according to the above-mentioned calculations automatically fulfilled by setting the diffraction conditions to an accurate two beam case. The influence of a deviation from the ideal two beam case is discussed in [Rub07b] to become larger for thicker specimens. In the following chapter 2.4 it will be shown that for EMCD experiments, thin specimens are required without exception. Thus, a small aberration of the ideal two beam case will have negligible influence on the measured EMCD signal.

The two Bloch waves were chosen to have slightly different wave vectors (equations 2.18 and 2.19) to receive the forward scattered beam and the diffracted beam as beats of the two Bloch waves. The second result is consistently the periodicity of spatial beats along the $\vec{z}$ direction with a period of $\frac{2\pi}{(\gamma^{(1)}-\gamma^{(2)})}$. As can be seen in the equations 2.29 and 2.30, when the electron beam enters the specimen ($z = 0$), all the intensity is located in the forward scattered beam ($\cos 0 = 1$ and $i \sin 0 = 0$). Within the specimen, the intensity is oscillating between forward scattered beam and diffracted beam with the periodicity of the beats, or - in other words - the total intensity is split between forward scattered beam and diffracted beam, depending on the thickness of the specimen. This effect is known as *Pendellösung* effect. As the dichroic effect is measured in the diffraction plane, it is thus foreseeable that the strength of the dichroic signal will also highly depend on the specimen thickness which will also be shown in the following chapter 2.4.

2.4. Calculations on EMCD

Although the interaction between the electron and the atom, leading to the dichroic effect, is not conceived in detail yet, it is possible to predict the strength of the dichroic signal using the dynamical diffraction theory and ab initio calculations of the electronic structure using the formalism of density functional theory [Rus07], [Rus08a], [Rub07a].

Within the ChiralTEM project, the groups of Pavel Novak and Jan Rusz (Prague) and the group of Peter Schattschneider (Vienna) modified the Wien2k [Wie2k] code, a software for electronic structure simulations to calculate the dependency of the dichroic effect of different parameters such as the element, the crystal structure and orientation, the specimen thickness or the exact position of the detector in the diffraction plane. These calculations are not part of this work and the Ansatz can be found in detail in [Rus07]. In contrast, some of the results, especially the interrelationship between specimen thickness and predicted dichroic effect, are essential for this work - especially for the specimen preparation (chapter 5) and for the EMCD experiment (chapter 6). Thus it is necessary to discuss these results here without going into depths of their theoretic background.

Using the definition

$$\sigma_{\text{rel}} = \left| \frac{\sigma^{+} - \sigma^{-}}{\sigma^{+} + \sigma^{-}} \right| \tag{2.31}$$

for the relative dichroic signal σ_{rel} with signals σ^{+} and σ^{-} for the different *helicities*, the graphs in figure 2.5 show the thickness dependency of the relative dichroic signal at the L_3-edges for fcc nickel in [001] orientation, bcc iron in [001] orientation and hcp cobalt in [0001] orientation.

In diffraction mode, the intensity of a certain Bragg spot highly depends on the specimen thickness as derived from dynamical diffraction theory in section 2.3. As the EMCD measurement is performed in diffraction mode, also the achievable dichroic signal depends on the specimen thickness. A thickness variation of only 10 nm can change the predicted dichroic signal from a high value to almost zero.

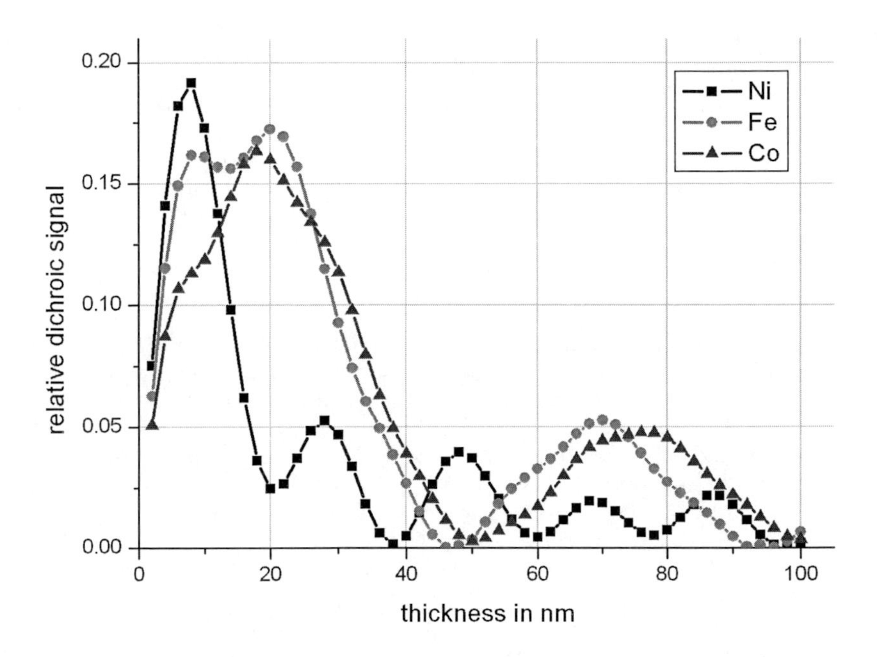

Fig. 2.5.: *Predicted relative dichroic signal at the L_3-edges of nickel ([001]-orientation, G-spot is (200)), iron ([001]-orientation, G-spot is (200)) and cobalt ([0001]-orientation, G-spot is $(11\overline{2}0)$) for an acceleration voltage of 200 kV and the specimen's magnetization saturated parallel to the electron beam. Induced by the Pendellösung variations of the intensities of Bragg spots (compare equations 2.29 and 2.30), the signal highly depends on the specimen thickness. Data provided by Jan Rusz, formerly Prague, now Uppsala.*

For a better facility of inspection, a listing of the optimum specimen thickness t_{opt}, the maximum relative dichroic signal $\sigma_{\mathrm{rel}}^{\mathrm{max}}$ at t_{opt} and the thickness interval with a relative dichroic signal of more than 10% is derived from the graphs in figure 2.5 and collected in table 2.2.

Whereas nickel provides the largest dichroic signal with 19.1%, the optimum specimen thickness has its largest amount for iron ($t_{\mathrm{opt}}(\mathrm{Fe}) = 20\,\mathrm{nm}$). The largest specimen thickness with a relative dichroic signal of more than 10% is found for cobalt at a specimen thickness of 30 nm. For each of the elements, the dichroic signal vanishes almost

Element	t_{opt}	$\sigma_{\text{rel}}^{\text{max}}$	$t(\sigma_{\text{rel}} > 10\%)$
Ni	8 nm	19.1%	4 nm to 12 nm
Fe	20 nm	17.2%	4 nm to 28 nm
Co	18 nm	16.3%	6 nm to 30 nm

Table 2.2.: *According to the calculations shown in figure 2.5, the optimum specimen thickness t_{opt} and the predicted maximum relative dichroic signal at the L_3-edge σ_{rel}^{max} are shown. The last column shows the thickness range with a predicted relative dichroic signal of more than 10% (the calculations were done in thickness steps of 2 nm and for an acceleration voltage of 200 kV).*

completely for certain specimen thicknesses. These results show that it is not offhand possible to recommend one of the three materials for EMCD experiments. Accurate investigations of the magnetic properties and the possibilities of specimen preparation are an essential part of this work and can be found in the chapters 4 and 5.

These calculations were done for an acceleration voltage of 200 kV which is the highest acceleration voltage available with the Tecnai F20 TEM in Vienna which was primarily dedicated for the EMCD experiments within the ChiralTEM project. The Tecnai F30 TEM in Regensburg, which was equipped with an energy filter in a late phase of the ChiralTEM project, provides an acceleration voltage up to 300 kV. The influence of the acceleration voltage on the relative dichroic signal can be seen rudimentarily in [Rus08a]. In brief, a change of the acceleration voltage from 200 kV to 300 kV leads to a stretching of the t-σ_{rel} curves. The optimum specimen thickness t_{opt} is thus shifted to a larger value. The shift is in the range of 10% and consequently also the thickness range with a predicted relative dichroic signal of more than 10% is enlarged. Up to now, no detailed calculations for either iron, nickel or cobalt, are available for other acceleration voltages than 200 kV. After this abridgment of the theoretic framework of EMCD, the next step towards an experimental verification has to be a definition of the experimental setup.

3. Experimental setup

In this chapter, the experimental setup which is required for the recording of dichroic spectra is briefly described. Beginning with a short review of transmission electron microscopy and the function of a post column energy filter, also two concrete measuring techniques for EMCD are discussed.

3.1. The Transmission Electron Microscope (TEM)

The Transmission Electron Microscope is not only the tool for the measurement of EMCD. It is also useful for specimen characterization. In this work, the TEM was used for the investigation of the crystal structure and for a confirmation of the ferromagnetic property of the provided specimen, reverting to Lorentz microscopy. Therefore, the different operating states *selected-area imaging mode* (observing the thickness contrast of a specimen), *diffraction mode* (observing diffraction patterns of a crystalline specimen) and *Lorentz mode* (observing magnetic domains) are shortly discussed in this section.

In the available FEI Tecnai F30, the electron beam is generated by a field emission gun (FEG). This technique has - in comparison to (exclusively) thermal emitters the advantage of a high spatial coherence of the beam which is important for instance for electron holography (used in section 7.2.2) or general interference experiments (compare section 8). The acceleration voltage of the electrons is adjustable in a range of 50 kV to 300 kV. After the acceleration, a condenser system generates a demagnified image of the emitter tip and provides a parallel illumination of the specimen. Beginning with the specimen plane, the image forming system is fundamentally different for the three discussed operation modes. The differences can be seen in figure 3.1. The further elements of the TEM, the magnifying system, the energy filter (optional) and the image aquisition by a CCD sensor is adapted to the particular requirements but their function remains unchanged in the different cases. In general, electron microscopes use magnetic lenses as described in detail in chapter 7.

TEM in imaging mode The conventional *selected area (SA) imaging mode* is used for overview screens of the investigated specimens. In this mode, the image plane of the objective lens is situated in the plane of the SA-aperture. Thus, this aperture can be used for the selection of a specimen area for further investigations (for example energy filtered transmission electron microscopy). The preimage plane of the diffraction lens conjoins with the image plane of the objective lens - an image of the specimen is transfered to the following magnifying system of the TEM [Rei97]. The diffraction angle can be limited

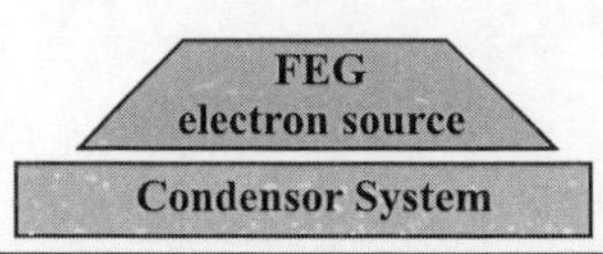

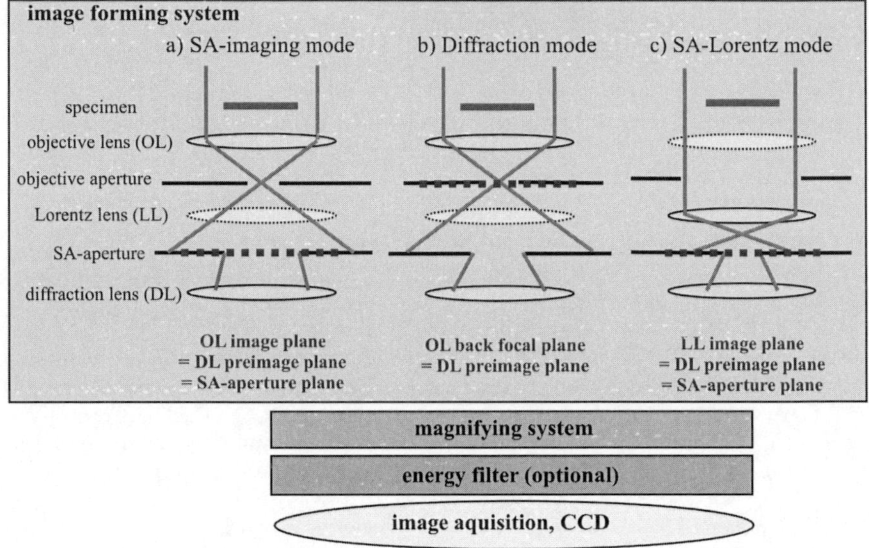

Fig. 3.1.: *Schematic setup of a TEM in SA (selected area) imaging mode (left side), Diffraction mode (center) and Lorentz-SA mode (right side). The essential difference occurs in the image forming system, consisting of specimen, objective lens and aperture, Lorentz lens, SA aperture and a combination of diffraction lens and first intermediate lens (in the following summarized to the diffraction lens) whereas the beam generation (electron source and condenser system) and the reproducing system (magnifying system, energy filter and image aquisition) remain basically unchanged. Inactive lenses are displayed with dashed lines.*
In the SA-imaging mode, the SA aperture is in the image plane of the objective lens and can thus select the observed area of the specimen. The diffraction lens grabs an image from the SA plane.
In diffraction mode, the diffraction lens grabs the diffraction pattern from the back focal plane of the objective lens. The objective aperture is used for a selection of different diffraction angles.
In Lorentz mode, the objective lens is replaced by the Lorentz lens with a longer focal distance. Thus, the specimen plane is free of magnetic fields.

by the objective aperture - different crystal orientations appear at different brightness and thus an evaluation of the crystallite sizes of a polycrystalline specimen is possible as shown in [Hus06].

TEM in diffraction mode In *diffraction mode*, the preimage plane of the diffraction lens conjoins with the back focal plane of the objective lens. Thus, no image of the specimen is transfered to the magnifying system, but a diffraction pattern of the specimen [Bru02]. Each position in the diffraction plane stands for a certain momentum transfer of the probe electron relative to the zero-beam (the unscattered component). Thus, the diffraction mode is an ideal tool for an EMCD measurement as the conditions concerning the momentum transfer (compare chapter 2) can be fulfilled easily by selecting momentum transfers with the spectrometer entrance aperture as described in detail in section 3.3.

TEM in Lorentz mode For investigations of magnetic specimens (as shown in chapter 5) it is often necessary to preserve the specimen from the magnetic field of the objective lens system[1] which would take influence on the domain structure of the specimen or even saturate the specimen completely, perpendicular to the specimen's plane. Thus, a TEM can be equipped with an alternative lens, called Lorentz[2] lens, which is farther away from the specimen plane as the objective lens [Zim95]. Because of the larger distance to the specimen which leads to a smaller refraction power (for which a smaller magnetic field is required), the magnetic influence on the specimen is negligible - at the expense of an inferior spatial resolution (about 2 nm in comparison to less than 0.2 nm in conventional imaging mode).

Having a specimen with magnetic domains (the domain formation is discussed in chapter 4), the electron beam is deflected by the component of the magnetization perpendicular to the electron's path by the Lorentz force (compare figure 3.2)

$$\vec{F_L} = e\left(\vec{v} \times \vec{B}\right) \tag{3.1}$$

with electron charge $e = -1,602 \cdot 10^{-19}\,\mathrm{C}$, the vector of movement $\vec{v}$ and the vector of the magnetic induction $\vec{B}$.

As the objective lens, also the Lorentz lens images the plane that it is focused on. If the Lorentz lens is set to *overfocus*[3], a plane below the specimen plane is focused - with a contrast generated by the magnetic deflection of the electron beam. In *underfocus*, the electron beams below the specimen plane must be virtually extended to the focus plane above the specimen plane - the magnetic contrast is reversed. Thus, the magnetic contrast of a Lorentz micrograph is reversed when changing from *underfocus* to *overfocus*. This imaging technique is called Fresnel imaging.

---

[1]the setup of the objective lens system of a TEM will be discussed in detail in chapter 7.

[2]Hendrik Antoon Lorentz (* 18 July, 1853, † 4 February, 1928)

[3]in *underfocus*, the lens current is *smaller* than in the focused case and thus the image plane of the lens is above the specimen. Contrariwise for *overfocus*.

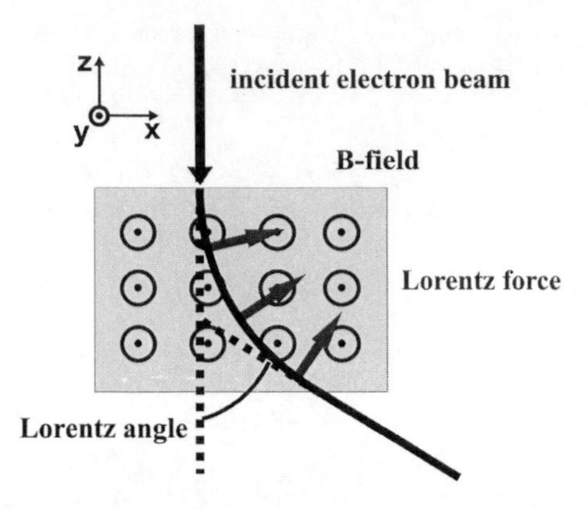

Fig. 3.2.: *In the field of a magnetic specimen, the incident electron beam is deflected to the Lorentz angle by the Lorentz force. According to figure 3.3, this deflection can generate a magnetic contrast in Lorentz micrographs.*

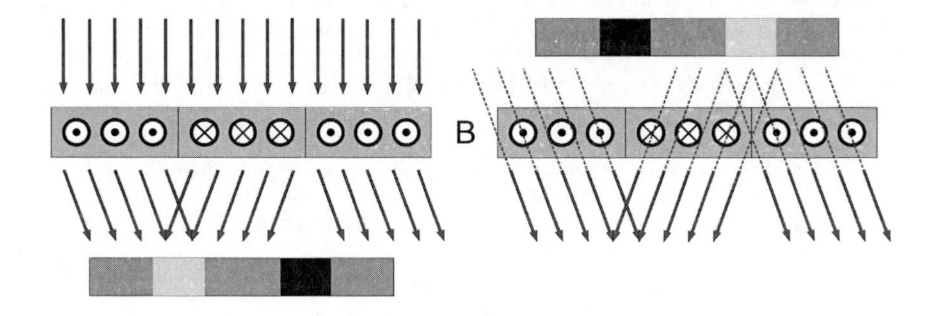

Fig. 3.3.: *Principle of Fresnel imaging: (left side) The incident electron beam is deflected due to the in-plane component of the specimen's magnetization. If the Lorentz lens has it's focus in a plane below the specimen plane (overfocus), the domain borders become visible as bright or dark lines. Setting the objective lens to underfocus (right side), the contrast is reversed as the beam seems to come along the virtual elongation of the deflected beams above the specimen.*

3.2. Electron Energy Loss Spectrocopy (EELS)

The available TEM is equipped with a post column energy filter[4] which is mounted below the viewing screen[5]. The functionality can be seen in figure 3.4. The polychrome (due to electron-specimen interaction, some electrons suffer an energy loss - compare section 2.1.2) electron beam is split in a self focusing energy dispersive sector magnet [Ege96]. The energy resolved spectrum[6] is recorded with a CCD sensor - one axis remains thereby a spatial axis, while the other axis becomes the energy loss axis (this fact will be used in section 3.3). To preserve the optical quality of the magnifying system (which is adjusted to an acceleration voltage of 300 kV), the acceleration voltage is increased by the desired energy loss of the investigated edges. This way, the energy of the relevant electrons remains unchanged below the specimen.

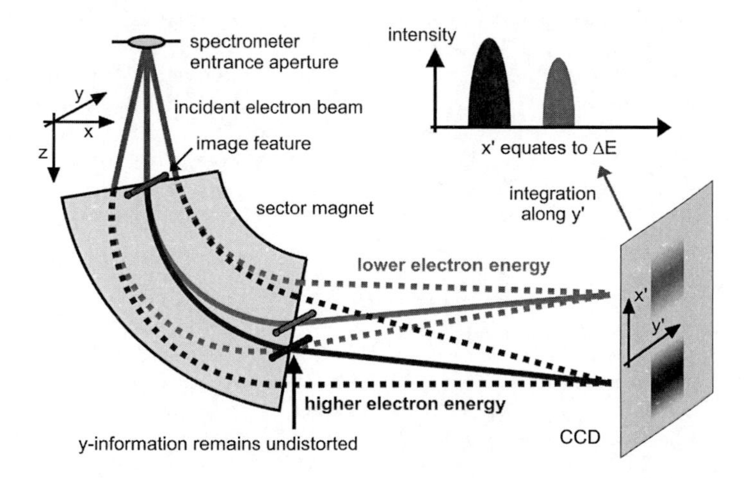

Fig. 3.4.: *Simplified sketch of the functionality of a post-column energy filter. Electrons that pass the spectrometer entrance aperture are deflected in the field of a sector magnet. In the focal plane of the sector magnet, the energy loss is selected by a slit aperture. Whereas the origin x-direction becomes the direction of the energy loss due to the energy splitting property of the magnetic field, the spatial information in y-direction is preserved to the CCD. An integration along the y' direction results in a noise reduced energy loss spectrum along x'.*

[4]GIF (Gatan Imaging Filter) Tridiem

[5]The technique of electron energy loss spectrometry was developed in the mid 1940s by Hillier and Baker [Hil43], [Hil44]

[6]A filtered imaging technique is also possible but it was not used in this work. Details can be found in [Ege96].

3.3. Setups for EMCD measurements

For the recording of the dichroic effect, two different techniques are available[7] - the *diffraction shift* method and the *spectrum spread* method. Both methods are presented in this section and the advantages and disadvantages are discussed. Whereas the *diffraction shift* method was used for the dichroic measurement in chapter 6, the commutation measurements in chapter 7 were done by using the *spectrum spread* method.

3.3.1. Diffraction shift method

Assuming ideal specimens (compare chapter 5.1), the experimental setup for EMCD must fulfill all the conditions listed in section 2.1. In particular, the measured electron has to suffer concurrently two different, perpendicular momentum transfers $\vec{q}_1$ and $\vec{q}_2$. In the diffraction mode, this condition is fulfilled by the interference of two partial electron waves with two Bragg spots (0 and G in the two beam case) as points of origin at any position on a Thales circle through the two Bragg spots (figure 3.5). The absolute value of the vector product $\vec{q}_1 \times \vec{q}_2$ is maximized at the intersection of the median line between 0 and G with the Thales circle (drawn as *position A* and *position B* in figure 3.5). Additionally, the phase shift between the partial waves $\vec{k}_1$ and $\vec{k}_2$ must be $\frac{\pi}{2}$. This is automatically fulfilled in a two beam case, compare section 2.3. If no ideal two beam case is adjusted, the phase shift diverges from $\frac{\pi}{2}$. This deviation is shown in [Rub07b] to be noncritical for small aberrations of less than 10% of the ideal two-beam case Laue center circle.

The diffraction pattern can be shifted relative to the (fixed) spectrometer entrance aperture using the *diffraction shift* functionality of the TEM to have either position A or position B in the spectrometer entrance aperture (compare figure 3.5)[8]. Thus, no variation of illumination or imaging parameters is required between the two measurements. According to theory (section 2.1) the difference between two spectra taken at the positions A and B show a dichroic difference if the specimen's magnetization is parallel to the electron beam.

3.3.2. Spectrum spread method

The *spectrum spread* method works different to the previously shown *diffraction shift* method. It uses the raw data of the energy filter CCD instead of the energy loss spectrum. As, after the sector magnet, the electron pattern on the CCD still contains spatial information in one direction (compare figure 3.4), the diffraction pattern can be adjusted relative by the spectrometer entrance aperture (respectively relative to the CCD as shown in figure 3.6) to have the 0-G axis in the direction of the energy loss (x-axis respectively x'-axis). Thus, position A and B are separated in $y = y'$ direction

[7] In the ChiralTEM project, also the possibility of a filtered imaging setup was discussed, but today, the signal intensity is several magnitudes too small to be used (see [Rub07a] for details).

[8] The exact location of position A and B can easily be calculated using the diffraction pattern "measuring" tool of the Tecnai user interface.

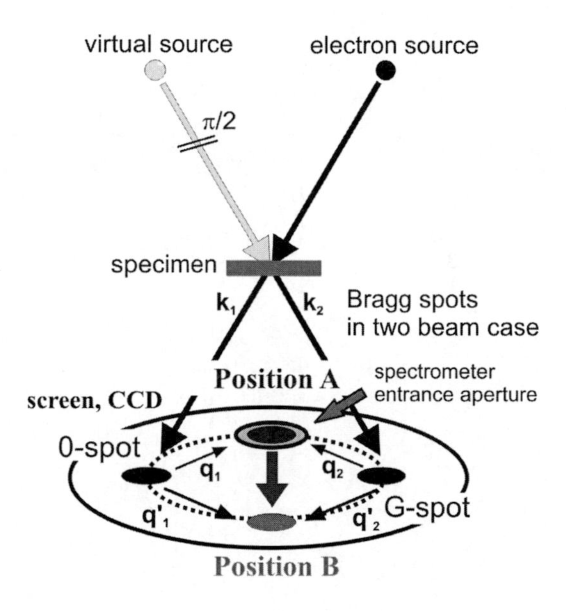

Fig. 3.5.: *Sketch of the diffraction shift method. The two points A and B on the Thales circle through the two Bragg spots are selected by a shift of the diffraction pattern relative to the spectrometer entrance aperture. At the positions A and B, the electrons coming from 0 and G interfere with suitable momentum transfers $\vec{q}_1$ and $\vec{q}_2$ respectively $\vec{q}_1{}'$ and $\vec{q}_2{}'$. The phase shift of $\frac{\pi}{2}$ between the partial waves $\vec{k}_1$ and $\vec{k}_2$ is present due to the two beam case (compare section 2.3).*

from the energy loss signal generated by the two Bragg spots. Two linescans at the y' coordinates corresponding to position A and B create two energy loss spectra containing the same dichroic difference as received with the *diffraction shift* method.

The advantages of the *spectrum spread* method are on the one hand an improved signal to noise ratio at the cost of an ideal dichroic signal, because not only electrons directly from the positions A and B count to the signal but also electrons from other positions in reciprocal space (and thus $\vec{q}_1$ not perpendicular to $\vec{q}_2$). On the other hand, only one measurement is required to get the two spectra. Considering the aquisition time, this advantage is only marginal as the low intensity along the linescans A and B requires further on an aquisition time of several 10 seconds similar to the diffraction shift method.

Disadvantages of the *spectrum spread* method are primarily the difficult adjustment of the diffraction pattern relative to the energy filter CCD. The diffraction pattern can be

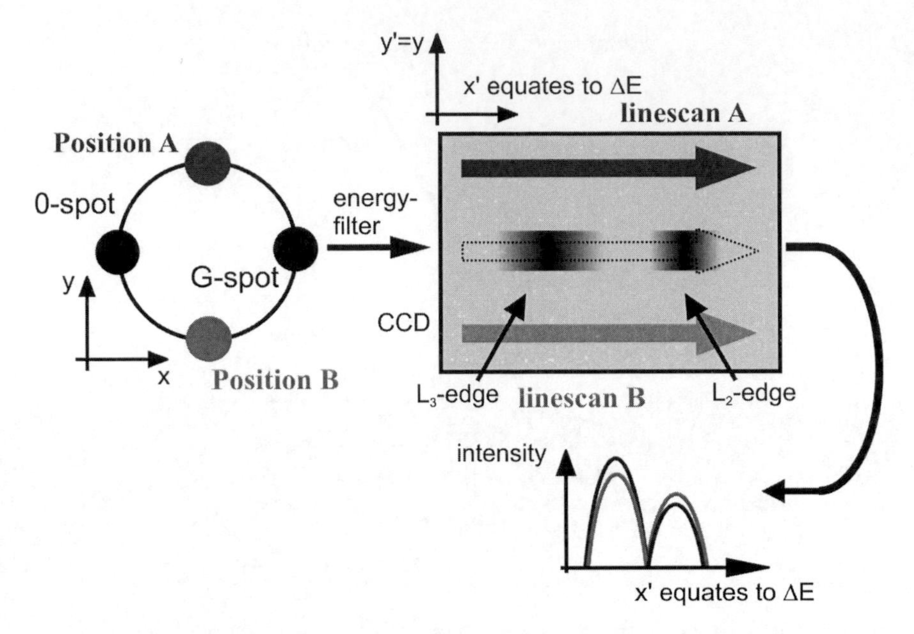

Fig. 3.6.: *In the spectrum spread method, the conservation of the spatial resolution along the y-direction after the energy filter is utilised (compare figure 3.4). The diffraction pattern is orientated with the 0-G-axis in the energy loss direction (x resp. x′). The linescan producing a common energy loss spectrum is shown by the dashed arrow. Making two symmetric linescans with the same distance to the 0-G-axis as the points A and B (blue and green arrow), the two dichroic spectra (bottom) are received (in analogy to figure 3.5).*

rotated by adjusting the projection lens P2 in the free lens control[9] but this rotation also causes a change of the effective camera length that has to be considered for the position of the two linescans. Due to these adjustments, the beam time on the specimen and thus an eventual beam damage (see section 6.2) is comparable to the *diffraction shift* method.

The experimental execution of the *spectrum spread* method can be improved by a separate alignment which expands the pattern on the CCD along the y'-axis. Details on this alignment can be found in appendix D.

[9]... or alternatively the TEMspy interface.

4. Magnetic considerations

In this chapter, details on the magnetization of a ferromagnetic specimen in a TEM are given. First of all, the strength of the magnetic fields which stem from the residual stray fields of the magnetic lenses at the specimen's location is discussed (section 4.1) and possible problems are demonstrated (section 4.2). Finally, the consequences are evaluated using calculations on the free energy (section 4.3) in a ferromagnetic specimen and micromagnetic simulations (section 4.4).

4.1. Ferromagnetic specimens in a TEM

Due to the exchange-coupling between adjacent electron spins, they try to minimize their (exchange-)energy by achieving parallel orientation. Thus, areas with a locally homogeneous magnetization $\vec{M}(\vec{r})$ can be observed in a ferromagnetic specimen. These areas are called magnetic domains. The behavior of these domains concerning different elements, specimen dimensions and external magnetic fields can be understood if considering also other contributions to the free energy of the ferromagnet as shown in section 4.3. Fundamentally, the presence of an external magnetic field enlarges areas with a local magnetization parallel to the external field at cost of the size of areas magnetized antiparallel to the external field. At a certain strength of the external field, the so called saturation field, all spins are oriented along this field.

In a transmission electron microscope, the specimen is usually[1] penetrated by the magnetic field of the objective lens. To get information about the absolute value of this lens' field, measurements with a Hall sensor specimen holder were performed. In this setup, a Hall sensor is mounted in a TEM specimen holder at the specimen's position [Ott01]. The Hall sensor (type KSY44, Infineon) is driven by a constant current source (7 mA) and generates an output voltage proportional to the magnitude of the magnetic field which can be measured with a multimeter (see figure 4.1 for experimental setup).

The conversion from Hall voltage to the magnetic field was done accordingly to the calibration of [Ott01]. According to the datasheet of the manufacturer, the linearity of the Hall sensor is better than $\pm 0.7\%$ up to a field of 1 T. As the aberration above 1 T is unknown, the error of the measurements is estimated generously to $\pm 5\%$. The Hall sensor can only detect the component of the field that is perpendicular to the specimen plane. Components in other directions (if existing) remain undetected.

[1]Except of operating in Lorentz mode, compare section 3.1 and section 8.3.

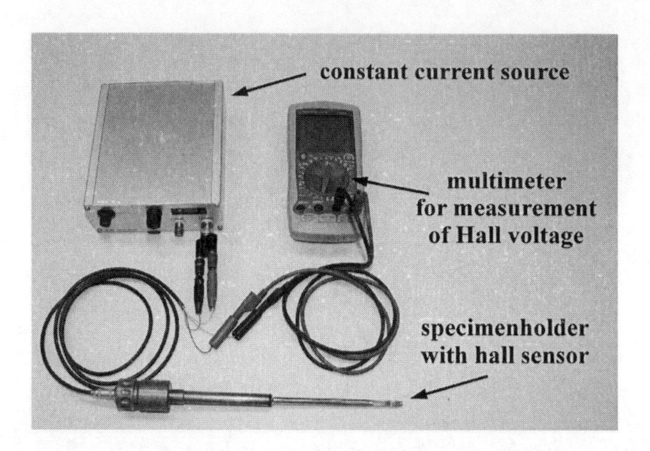

Fig. 4.1.: *Setup for the measurement of the magnetic field at the specimen's location in the TEM: a Hall sensor in the tip of the specimen holder is supplied by a constant current source. The generated output Voltage is proportional to the magnetic field and can be measured with a multimeter.*

Fundamentally, the magnetic field of a magnetic electron lens depends on the acceleration voltage. The higher the acceleration voltage, the higher is the (relativistic) mass of the electrons. Additionally, the retention time in the magnetic field is reduced. Thus, a stronger field is required to achieve the same deflection angle of the electron beam. To have comparable values, the objective lens excitation is denoted by a per cent value of the maximum lens current at a specific acceleration voltage. At different acceleration voltages, the same per cent value thus indicates the same refraction power, but implies a different lens current and thus a different magnetic field at the specimen's location.

First, the magnetic field versus the acceleration voltage was measured in a range from 50 kV (the lowest acceleration voltage at the Regensburg Tecnai F30) to 300 kV (the highest available acceleration voltage). In each case, the objective lens was excited to the maximum possible lens current at the given acceleration voltage[2]. The results are shown in figure 4.2. The field range starts at $1.04T \pm 5\%$ at an acceleration voltage of 50 kV and reaches $2.04\,\mathrm{T} \pm 5\%$ at an acceleration voltage of 300 kV.

Additionally, the magnetic field was measured versus the objective lens excitation (in per cent of the maximum possible lens current at a given acceleration voltage) for acceleration voltages of $100kV$, $200kV$ (the maximum available acceleration voltage at

[2]For each acceleration voltage, the lens excitation can be adjusted in a range from "0%" to "100%". For different acceleration voltages, the same per cent values of excitation require different lens currents to achieve the same beam deflection. At "full" lens excitation (100%) the lens current varies from 7.2 A at 100 kV to 13.7 A at 300 kV.

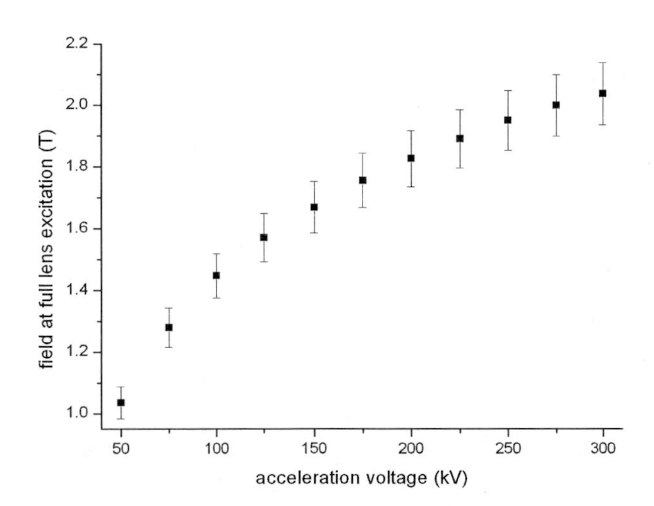

Fig. 4.2.: *The lens current - and thus also the magnetic field of the lens - depends on the acceleration voltage of the electron beam. The graph shows the magnetic field $\mu_0 H$ of the (for each acceleration voltage fully excited) objective lens at the specimen's location for different acceleration voltages.*

the Vienna Tecnai F20, the TEM designated for EMCD measurements in the ChiralTEM project) and $300kV$ (see figure 4.3). Whereas the graph is approximately linear up to an acceleration voltage of $100kV$, the beginning saturation of the pole pieces becomes noticeable by a flattening of the curves at high excitations at higher voltages.

Usually, TEM measurements are done at an objective lens excitation[3] of roughly 90%. The magnetic fields at this excitation can be read off table 4.1.

acc. voltage	magnetic field
100 kV	$1.33\,\mathrm{T} \pm 5\%$
200 kV	$1.73\,\mathrm{T} \pm 5\%$
300 kV	$1.96\,\mathrm{T} \pm 5\%$

Table 4.1.: *Magnetic field $\mu_0 H$ of the objective lens (excited to 90% of the maximum possible current at a given acceleration voltage) at the specimen's position.*

[3]The objective lens excitation defines the defocus of the specimen plane.

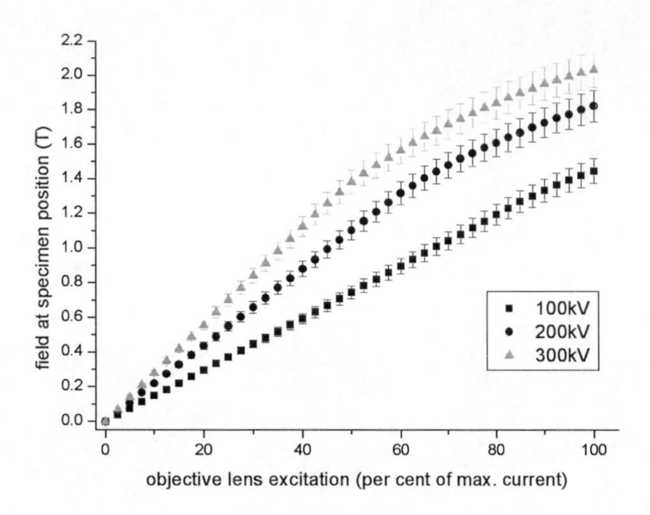

Fig. 4.3.: *Magnetic field $\mu_0 H$ of the objective lens at the position of the specimen versus the objective lens excitation in per cent of the maximum possible excitation for three acceleration voltages (*$100\,kV$*,* $200\,kV$ *and* $300\,kV$*).*

4.2. Motivation for magnetic calculations

For the largest possible dichroic signal, it is necessary that the magnetization of the specimen is parallel (or antiparallel) to the electron beam (see chapter 2). Due to the magnetic field of the objective lens (compare previous section 4.1), the specimen's magnetization is forced to fulfill this precondition, but micrographs of a cobalt specimen taken at a large underfocus[4] (see chapter 3.1) show the existence of remaining magnetic domains in the field of the objective lens of $1.6T$ (compare figure 4.4). The magnetization in these domains has a different orientation than the magnetization in the rest of the specimen - if one of these domains is partially in the area illuminated by the electron beam during a dichroic measurement, the received dichroic signal would be drastically reduced. Thus it must be one part of the characterization of specimens for the ChiralTEM project to investigate the degree of magnetic saturation of the provided specimens.

[4]Underfocussing the objective lens allows on the one hand the visualization of magnetic domains (see section 3.1) and reduces on the other hand the field of the objective lens.

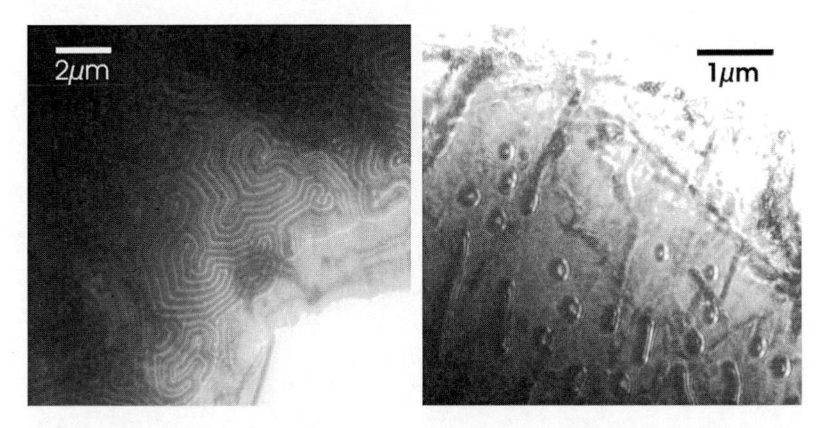

Fig. 4.4.: *Magnetic domains in a cobalt specimen, taken with a Philips CM30. Left side: Lorentz micrograph of maze domains without external field. Right side: normal imaging mode with strongly underfocused objective lens. Even in an external field of about* $1.6T$*, parts of the magnetic domains resist the external field. The area between the remaining domains is saturated in the direction of the external field.*

4.3. Calculations on the free energy

A first impression of the magnetic situation in ferromagnetic specimens can be derived by calculating the free energy E_f [Wol04],

$$E_f = E_E + E_Z + E_D + E_A, \tag{4.1}$$

with exchange energy E_E, Zeeman energy E_Z, dipole- or stray field energy E_D and anisotropy energy E_A. These contributions to the free energy are discussed in the following lines.

The local magnetization $\vec{M}(\vec{r})$ with its saturation magnetization M_S as absolute value can be expressed by a vector field $\vec{m}(\vec{r})$,

$$\vec{M}(\vec{r}) = M_S\,\vec{m}(\vec{r}), \tag{4.2}$$

with a boundary condition $(\vec{m}(\vec{r}))^2 = 1$. As nature always wants to minimize the free energy, most of the reactions of the local magnetization $\vec{M}(\vec{r})$ on a change of a certain parameter can be understood.

- **Exchange energy**
  The exchange energy

$$E_E = A \cdot \int_V \left(\vec{\nabla}\vec{m}(\vec{r})\right)^2 \mathrm{d}V, \tag{4.3}$$

  with the exchange constant[5] A and the volume of the specimen V is responsible for the spontaneous magnetization of a ferromagnetic material. It is minimized if all magnetic spins have the same orientation and thus forces the specimen to an homogeneous magnetization without magnetic domains [Lan35].

- **Zeeman energy**
  The Zeeman energy describes the interaction of the magnetization $\vec{M}(\vec{r})$ with an external magnetic field $\vec{H}(\vec{r})$

$$E_Z = -\mu_0 \int_V \mathrm{d}V \vec{M}(\vec{r}) \cdot \vec{H}(\vec{r}), \tag{4.4}$$

  with the magnetic field constant $\mu_0 = 1.257 \cdot 10^{-6}\,\frac{\mathrm{Vs}}{\mathrm{Am}}$. The Zeeman energy is minimized if the magnetization is orientated parallel to the external field and thus the magnetization always wants to align with an external field.

- **Dipole energy**
  If the magnetization of a specimen is not parallel to the specimen's surfaces or boundaries, a magnetic stray field leaks from the specimen. This stray field - also called demagnetizing field $\vec{H}_D$ - caused by the magnetization $\vec{M}(\vec{r})$ can be calculated as [Hoe04]

$$\vec{H_D}(\vec{r'}) = -M_S \,\mathrm{grad} \left(\int_V \mathrm{d}V \frac{-\vec{\nabla}_{\vec{r}} \cdot \vec{m}(\vec{r})}{\left|\vec{r} - \vec{r'}\right|} + \int_S \mathrm{d}S \frac{\vec{m}(\vec{r}) \cdot \vec{n}(\vec{r})}{\left|\vec{r} - \vec{r'}\right|} \right). \tag{4.5}$$

  Here, S is the surface of the specimen and $\vec{n}(\vec{r})$ is the surface normal vector at the position $\vec{r}$. The term $\vec{\nabla} \cdot \vec{m}(\vec{r})$ can be regarded as volume charges and $\vec{m}(\vec{r}) \cdot \vec{n}(\vec{r})$ can be regarded as surface charges of the magnetization. The energy of such a leaking field is called stray field energy or dipole energy [[Bro62]]

$$E_D = -\frac{1}{2}\mu_0 M_S \int_V \mathrm{d}V \vec{m}(\vec{r}) \cdot \vec{H}_D(\vec{r}). \tag{4.6}$$

  This contribution to the free energy wants to restrain stray fields by forcing the magnetization parallel to the specimen's surfaces and boundaries.

[5] examples for A can be found in table 4.3

- **Anisotropy energy**

  To realign the (randomly distributed) magnetic spins along the direction of an external magnetic field $\vec{H}$, the magnetization work W_M has to be spent, where

$$W_M = \mu_0 \int_0^{M_S} \vec{H} \cdot \mathrm{d}\vec{M}. \tag{4.7}$$

  If W_M depends on the direction of $\vec{H}$, the specimen has a so called magnetic anisotropy. Directions with minimum magnetization work are called easy axes and the other way round directions with maximum magnetization work are called hard axes. Uniaxial anisotropy is existent if the specimen has only one allocated axis - either easy or hard. For the simple case of uniaxial anisotropy[6], the anisotropy energy is given by

$$E_{A_{uni}} = V\, K_{\mathrm{uni}} \sin^2\theta \tag{4.8}$$

  with an anisotropy constant K_{uni}, an angle θ between the uniaxial axis, the magnetization $\vec{M}$ and the volume of the specimen V. The origin of this energy can be found in the crystalline structure: Crystal axes often are anisotropic axes due to an interaction between electron spins and the atomic angular momentum. A mechanical strain can cause magnetic anisotropy, too. At fixed parameters (specimen geometry, external field), the directions of anisotropic axes can force the magnetization of a specimen to be either in plane or out of plane.

The free energy can be used for a first estimation of the specimen's magnetization [Schn02]. The results of the calculations should be a value for the out of plane component of the magnetization at a given perpendicular external magnetic field in z-direction,

$$\vec{H} = H_0 \begin{pmatrix} 0 \\ 0 \\ 1 \end{pmatrix} \tag{4.9}$$

and to find the external field that is necessary to saturate the specimen completely parallel to the external field. The geometry is chosen as seen in figure 4.5: The external field only has a component in z-direction and the specimen plane is the x-y-plane. The specimen has a cylindrical shape with diameter d and thickness t. The diameter is assumed to be large compared to the thickness - the ratio[7]

$$d_t = \frac{d}{t} \tag{4.10}$$

is much larger than 1.

[6]Other examples can be found in [Hub98], but will not be used in this work.

[7]For a boundless film of infinite size, $\frac{1}{d_t}$ is always zero and the thickness of the film is irrelevant.

With an angle ϕ between magnetization and specimen plane, the magnetization is defined as

$$\vec{M} = M_S \begin{pmatrix} \cos\phi \\ 0 \\ \sin\phi \end{pmatrix}. \tag{4.11}$$

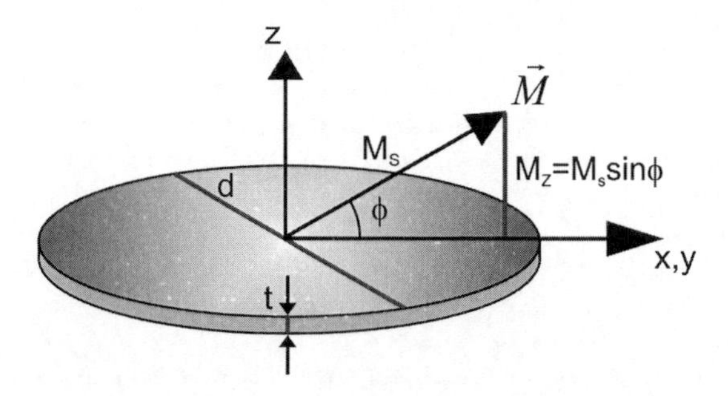

Fig. 4.5.: *sketch of a cylindrical specimen with diameter d, thickness t and an homogeneous magnetization $\vec{M}$, enclosing the angle ϕ with the specimen plane. The specimen is assumed to be thin, $d >> t$.*

The values for the thickness are chosen according to the predicted values for a large dichroic signal [see chapter 2.4], $t_{\mathrm{Fe}} = 20\,\mathrm{nm}$, $t_{\mathrm{Ni}} = 8\,\mathrm{nm}$ and $t_{\mathrm{Co}} = 18\,\mathrm{nm}$. The diameter d is chosen $1\,\mu\mathrm{m}$ and also the limit $d \to \infty$ is considered.

As a simplification, the specimen's magnetization is assumed to be homogeneous - all magnetic spins are oriented in the same direction. Therefore, the exchange energy (equation 4.3) becomes zero, $E_E = 0$. This presumption will only be appropriate near the saturation field which is the interesting area for the present calculations.

The magnetic anisotropy is only considered for the cobalt specimen (with uniaxial anisotropy) as the anisotropy energies for nickel and iron are about two orders of magnitude smaller than the corresponding stray field energy and thus negligible. The contributions to the free energy can thus be written as:

- **Exchange energy** (presumption: homogeneous magnetization):

$$E_E = 0. \tag{4.12}$$

- **Zeeman energy** (with equations 4.4 and 4.9):

$$E_Z = -\mu_0 \int \mathrm{d}V\, \vec{M}(\vec{r}) \cdot \vec{H}(\vec{r}) = -\mu_0 V M_S H_0 \sin\phi. \tag{4.13}$$

- **Stray-field energy** (with equation 4.6 and
$\vec{H}_D = -M_S \begin{pmatrix} \frac{\pi}{4d_t} & 0 & 0 \\ 0 & \frac{\pi}{4d_t} & 0 \\ 0 & 0 & 1-\frac{\pi}{2d_t} \end{pmatrix} \cdot \begin{pmatrix} \cos\phi \\ 0 \\ \sin\phi \end{pmatrix}$, [Hus06]):

$$\begin{aligned} E_D &= -\frac{1}{2}\mu_0 M_S \int_V \mathrm{d}V \vec{m}(\vec{r}) \cdot \vec{H}_D(\vec{r}) \\ &= \frac{1}{2}\mu_0 V M_S^2 \left(\frac{\pi}{4d_t}\cos^2\phi + \left(1 - \frac{\pi}{2d_t}\right)\sin^2\phi \right). \end{aligned} \tag{4.14}$$

- **Anisotropy energy** (with equation 4.8 and $\theta = 90° - \phi$ and thus $\sin\theta = \cos\phi$):

$$E_{A_{\text{uni}}} = V \cdot K_{\text{uni}} \cdot \cos^2\phi. \tag{4.15}$$

For a fixed ratio d_t and a given external field $B_0 = \mu_0 H_0$, the derivative of the free energy must be zero at an extremum of the free energy,

$$\frac{\mathrm{d}}{\mathrm{d}\phi} E_F = 0. \tag{4.16}$$

It can be shown [Hus06] that a minimum of the free energy can be found for

$$\phi = \arcsin \frac{4 M_S B_0}{\mu_0 M_S^2 \left(4 - 3\frac{\pi}{d_t}\right) - 8K_{\text{uni}}} \tag{4.17}$$

between the magnetization and the specimen plane. Therefore, according to figure 4.5, the z-component of the magnetization is given by

$$M_Z = M_S \cdot \sin\phi = \frac{4 M_S^2 B_0}{\mu_0 M_S^2 \left(4 - 3\frac{\pi}{d_t}\right) - 8K_{\text{uni}}}. \tag{4.18}$$

For a calculation of the saturation magnetization one can divide equation 4.18 by M_S and set $\frac{M_Z}{M_S} \equiv 1$. Thus, B_0 can be calculated to

$$B_0 = \mu_0 M_S \left(1 - \frac{3}{4}\frac{\pi}{d_t}\right) - 2\frac{K_{\text{uni}}}{M_S}. \tag{4.19}$$

The dependence of the M_Z component[8] on the external field is shown in figure 4.6 for iron ($t_{\text{Fe}} = 20\,\text{nm}$), nickel ($t_{\text{Ni}} = 8\,\text{nm}$), and cobalt ($t_{\text{Co}} = 18\,\text{nm}$) - the specimen thicknesses with the largest predicted dichroic signal (according to section 2.4) - on the one hand for a specimen diameter of $d = 1\,\mu\text{m}$ and on the other hand for the limit $d \to \infty$ which is equivalent to a infinite expanded layer specimen. By demanding $\sin\phi = 1$, the smallest external field to cause a complete saturation of the specimen can be calculated out of equation 4.19.

[8]Normalized by $\left|\frac{M_Z}{M_S}\right|$ to the interval from 0 to 1.

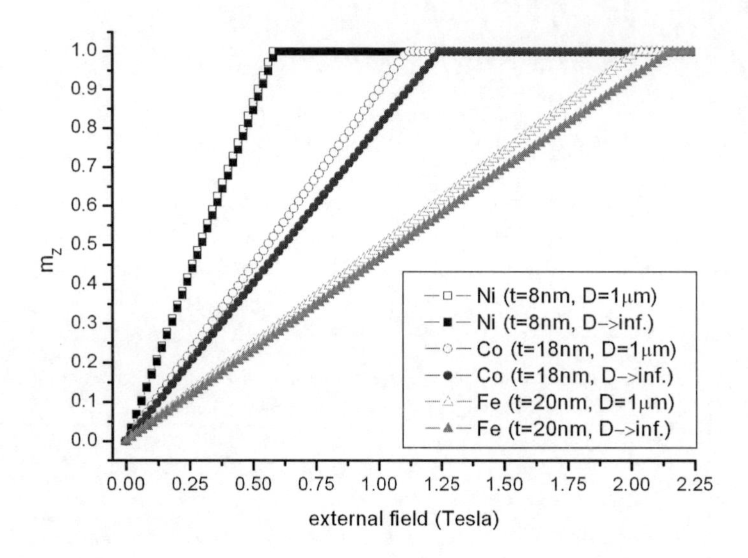

Fig. 4.6.: *Normalized perpendicular component* $m_Z = \left|\frac{M_Z}{M_S}\right|$ *of the magnetization versus the external magnetic field* $\mu_0 H_Z$ *for nickel, cobalt and iron cylindric films with a diameter* $d = 1\,\mu m$ *and the limit* $d \to \infty$ *according calculations on the free energy.*

Element	$\mu_0 H_{\text{ext}}(d = 1\,\mu\text{m})$	$\mu_0 H_{\text{ext}}(d \to \infty)$
Ni	0.59 T	0.60 T
Co	1.12 T	1.20 T
Fe	2.00 T	2.10 T

Table 4.2.: *For nickel, iron, and cobalt, the perpendicular, external fields that are necessary for a complete saturation of the specimen are calculated according to equation 4.18.*

The external fields necessary for a complete saturation are shown in table 4.2 to be (for $d \to \infty$) 0.60 T for a nickel, 1.20 T for a cobalt, and 2.10 T for an iron film. The saturation field of nickel is thus far below the magnetic field of the objective lens (1.3 T - 2 T, depending on the acceleration voltage - compare section 4.1). Any nickel specimen is expected to be completely saturated at the given experimental setup (see chapter 3).

Therefore, no further calculations or simulations are necessary for nickel specimens. According to the calculations, the saturation field for iron is larger than the maximum achievable external magnetic field in a TEM. Thus, iron specimens are expected to be not completely saturated in the direction of the electron beam. In contrast, the saturation field of cobalt is in the range of the objective lens' field. Furthermore, these estimations do not consider the appearance of magnetic domains as described in section 4.2.

To assure the results for iron and cobalt specimens it was decided to perform more detailed magnetic calculations, considering also the appearance of magnetic domains. These advanced calculations were done with a software for micromagnetic simulations and the results are shown in the following section.

4.4. Micromagnetic simulations

The Landau-Lifshitz-Gilbert (LLG) equation [Hub98]

$$\frac{\mathrm{d}}{\mathrm{d}t}\vec{m} = -\frac{|\gamma_0|}{1+\alpha^2}\left(\vec{m} \times \vec{H}_{\mathrm{eff}}\right) - \frac{\alpha\,|\gamma_0|}{1+\alpha^2}\left(\vec{m} \times \left(\vec{m} \times \vec{H}_{\mathrm{eff}}\right)\right) \tag{4.20}$$

is the equation of motion for a magnetic moment $\vec{m}$ with a gyromagnetic ratio γ_0 and a phenomenological damping parameter α [Hoe04]. After a change of the effective magnetic field $\vec{H}_{\mathrm{eff}}$, the magnetic moment precesses around this effective field and converges to its direction due to the damping. The effective field can be calculated from the external field, the exchange field and the stray field [Hin02].

To calculate the direction of the local magnetization, the specimen is divided into many cubic cells. To allow ferromagnetic coupling between adjacent cells, the edge length of the cells has to be smaller than the magnetic exchange length l_A of the simulated material,

$$l_{ex} = \sqrt{\frac{2A}{\mu_0 M_S^2}}. \tag{4.21}$$

The exchange constant A and the saturation magnetization M_S and thus the exchange length are specific to the ferromagnetic material and listed in table 4.3

For the simulations, the edge length is set to 3 nm for iron and 4 nm for cobalt. As the aspired specimen thicknesses are larger than the edge lengths, several layers are combined to a film using multilayer simulations (see figure 4.7 for a schematic view). Finally, in order to approach an infinite extended magnetic film, periodic boundary conditions are set.

With these specifications, equation 4.20 is being integrated iteratively until the change of the magnetization is below a predefined threshold in all cells. Depending on the threshold, the resulting magnetization is then a state of equilibrium. These calculations can be performed with different micromagnetic simulation packages. In this work, the package LLG [LLG08] was used because it is well adapted for multilayer simulations and the use

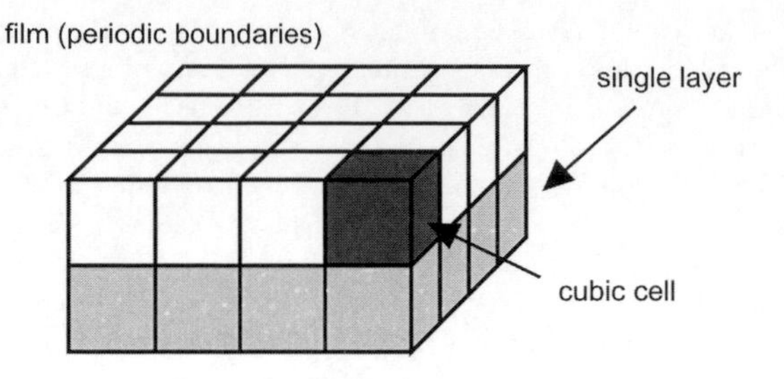

Fig. 4.7.: *Schematic view of the buildup of a film in multilayer micromagnetic simulations, consisting of different layers and cubic cells. In each cell, the direction of the magnetization is calculated by solving the LLG equation (equation 4.20) iteratively.*

Element	A	M_S	K	l_{ex}
Ni	$9 \cdot 10^{-12}\,\mathrm{J/m}$	$470 \cdot 10^3\,\mathrm{A/m}$	$-4.5 \cdot 10^3\,\mathrm{J/m^3}$	$7.7\,\mathrm{nm}$
Co	$30.5 \cdot 10^{-12}\,\mathrm{J/m}$	$1414 \cdot 10^3\,\mathrm{A/m}$	$40 \cdot 10^4\,\mathrm{J/m^3}$	$5.1\,\mathrm{nm}$
Fe	$21 \cdot 10^{-12}\,\mathrm{J/m}$	$1714 \cdot 10^3\,\mathrm{A/m}$	$47 \cdot 10^3\,\mathrm{J/m^3}$	$3.4\,\mathrm{nm}$

Table 4.3.: *Magnetic parameters for Fe, Ni and Co: Exchange constant A, saturation magnetization M_S, anisotropy constant K (uniaxial anisotropy for Co, kubic anisotropy for Ni and Fe) and the exchange length l_{ex} calculated according to equation 4.21.* These parameters were used for both, the free-energy calculation (section 4.3) and the micromagnetic simulations (present section).

of dual core processors[9] (see also [Per04] for details on LLG). The simulations disregard effects occurring due to a non-zero temperature, such as a shift of the saturation magnetization M_S or thermal excitation of magnetic moments in the specimen.

The simulation of a cobalt layer (figure 4.8) shows maze domains which are characteristic for materials with a strong uniaxial anisotropy perpendicular to the specimen's plane, such as (0001)-oriented cobalt or several yttrium-iron garnets (YIG). For this simulation, 12 layers, each with 256×256 cells with a random distribution of the magnetization direction in each cell as initial condition was relaxed without external field.

In all the simulated films, the perpendicular component of the magnetization is coded in blue (into the plane) and red (out of the plane). The arrows (wherever plotted) represent the direction and strength of the in-plane component. A comparison of a surface layer (layer 1 of 12) and a layer from the middle of the film (layer 6 of 12) shows a different ratio between in-plane on out-of-plane magnetization. For a reduction of the stray

[9]Nevertheless, each cobalt film requires a simulation time of several days.

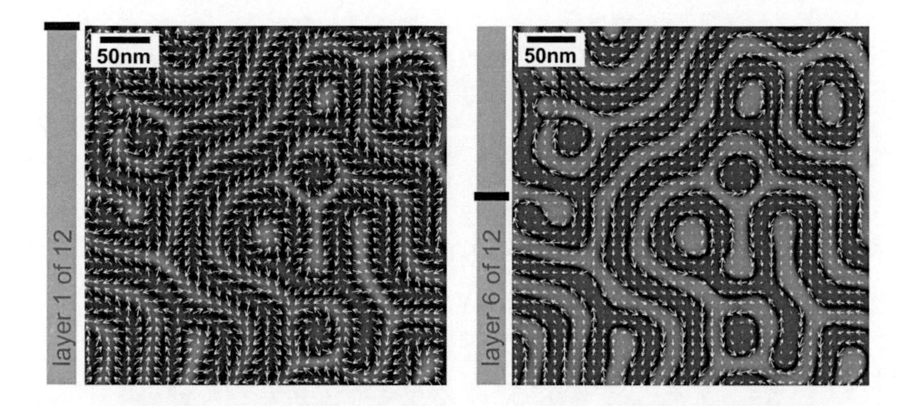

Fig. 4.8.: *LLG multilayer simulation of a cobalt film (t = 48 nm) with a size of 1024 nm × 1024 nm. The out-of-plane component of the magnetization is coded in dark grey (into the plane) and light grey (out-of-plane). The arrows represent the direction and strength of the in-plane component.*
Whereas in the middle of the film (right side, layer 6 of 12) the in-plane component of the magnetization is rather small (short yellow arrows and slim (black) walls between the domains), on the surfaces of the film (left side, layer 1 of 12) more magnetic moments are oriented in the specimen plane.

field energy (equation 4.6), the magnetization turns parallel to the specimen surface in the outer layers. This fact may be important when comparing future EMCD in Lorentz mode (non saturated specimens) with surface sensitive reflection XMCD, for example XMRM[10] (see [Tie08] or [Gei03] for details on XMRM).

Rising an external magnetic field perpendicular to the specimen plane, the areas with a magnetization parallel to the external field grow due to the Zeeman energy (equation 4.4) at the cost of the areas antiparallel to the external field. At a certain field (depending on the specimen thickness), the maze domains decompose into magnetic bubbles (compare section 5.3.1). Simulated slices through different depths of such a magnetic bubble are shown in figure 4.9. A schematic cross section of maze domains and a magnetic bubble is drawn in the figures 4.10 respectively 4.11. The generation of bubble domains out of maze domains with an increasing external field can be observed in figure 4.12.

[10]X-ray Resonant Magnetic Reflectometry

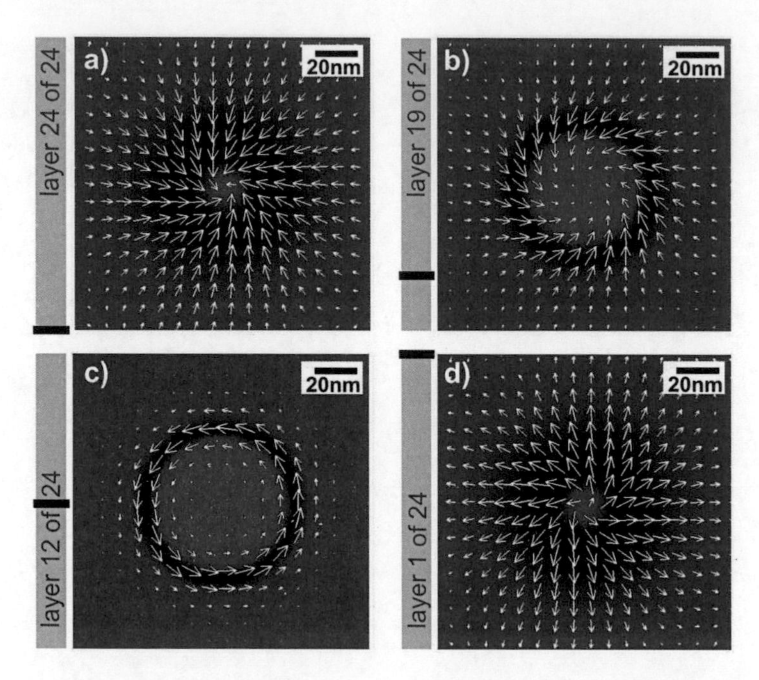

Fig. 4.9.: *LLG multilayer simulation of a magnetic bubble in a cobalt film ($t = 96\,nm$) at a perpendicular external field of* 1.0 *T. The 4 cuts through different layers (see the grey bars for the position of each layer in the film) show the characteristic change of the in plane component (yellow arrows) across the film. A cross section of a bubble is sketched in figure 4.11.*

Fig. 4.10.: *Schematic view of maze-domains in a specimen with a strong magnetic anisotropy perpendicular to the specimen plane (such as cobalt or yttrium-iron garnets)*

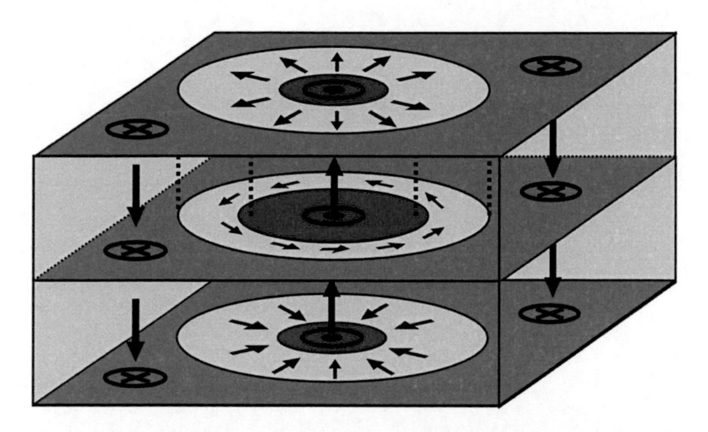

Fig. 4.11.: *Schematic view of a bubble domain according to the simulation shown in figure 4.9. Bubble domains grow out of maze domains with an increasing perpendicular external magnetic field.*

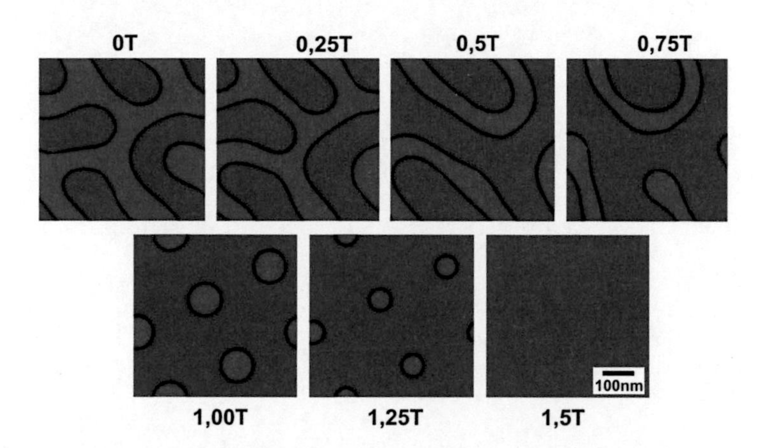

Fig. 4.12.: *LLG multilayer simulation of a cobalt film ($t = 96\,nm$) with periodic boundary conditions. The size of the displayed area is $512\,nm \times 512\,nm$ ($128 \times 128 \times 24$ cubic cells) each and the cuts were taken from the middle of the film (layer 12 of 24). Rising an external perpendicular magnetic field in the "red"-direction creates magnetic bubbles, starting at a field of about 1 T. Complete saturation is present at a field between 1.25 T and 1.5 T. The exact value of the saturation field can be determined by extrapolation (see graph 4.13).*

Using a simulation series such as the one shown in figure 4.12, the normalized perpendicular magnetization

$$m_Z = \frac{1}{M_S} \left| \frac{\sum_{\text{cells}} M_Z}{N_{\text{cells}}} \right| \tag{4.22}$$

can be calculated from the simulation data using a matlab [Mat08] script. Cobalt films with a thickness of 36 nm[11], 48 nm, 72 nm and 96 nm[12] were simulated. The saturation field is determined by an extrapolation of a linear fit of the data points below the saturation field ($M_Z = M_S$, $m_Z = 1$). The results are summarized in the following table 4.4 and the data points with linear fits are plotted in graph 4.13.

t(cobalt)	saturation field
36 nm	1.21 T
48 nm	1.23 T
72 nm	1.31 T
96 nm	1.41 T

Table 4.4.: *Saturation field $\mu_0 H$, depending on the thickness of the unbounded cobalt layer according to LLG simulations.*

The results of the free energy Ansatz (1.20 T for a boundless cobalt layer) correspond well with the simulation of the 36 nm layer (uniform magnetization). For a higher film thickness, magnetic maze domains appear and increase the saturation field noticeable. The experiment of section 4.2 which shows magnetic domains resisting an external field of 1.6 T can not be retraced with the simulations. But, on the one hand, the specimen thickness in the shown image is unknown and, on the other hand, a real specimen is different from an ideal, infinite film. There may be thickness variations and defects that can act as pinning centers for the domains. Thus, despite the simulation results, it is possible that sporadic bubble domains resist the field of the objective lens and they can - if illuminated by the electron beam in a dichroic measurement - reduce the dichroic signal.

Calculations on the free energy (chapter 4.3) result in a saturation field of 2.10 T for a boundless iron layer. As these calculations contain several simplifications, also this result is verified using micromagnetic simulations as shown above for the cobalt films. The results of the simulations are shown in table 4.5 and in graph 4.14. They confirm the independence of the saturation field of a boundless film from the film thickness as the

[11] At a thickness of $t = 36$ nm the film has homogeneous magnetization (no appearance of maze domains)

[12] Although a 96 nm cobalt film is too thick for dichroic experiments, this thickness was simulated in order to comprehend the situation shown in figure 4.4. Even at 96 nm, in the simulation no domains resist a field of more than 1.5 T. A simulation of more layers was not possible due to a limitation of computing time and main memory.

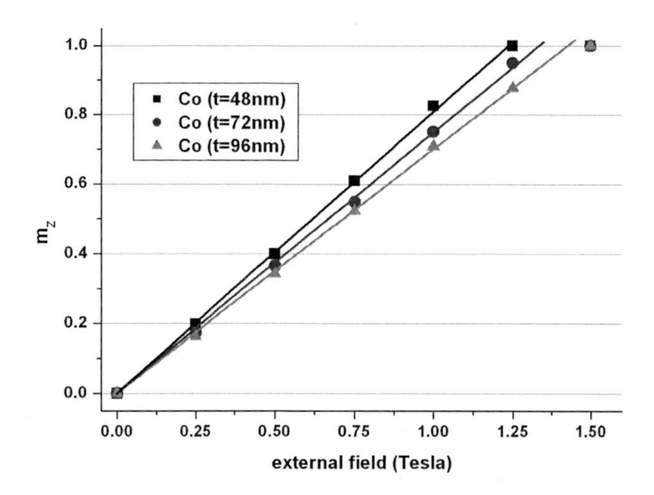

Fig. 4.13.: *Normalized perpendicular component* $m_Z = \left|\frac{M_Z}{M_S}\right|$ *of the magnetization in cobalt layers (48 nm, 72 nm and 96 nm) with periodic boundary conditions versus the external field* $\mu_0 H_Z$*, calculated with LLG. The linear fits (including only data points before saturation) show lines through origin with a gradient of* $0.81\,\frac{1}{T}$ *(at 48 nm),* $0.76\,\frac{1}{T}$ *(at 72 nm) and* $0.71\,\frac{1}{T}$ *(at 96 nm).*
The difference between a cobalt layer with $t = 48$ *nm and one with* $t = 36$ *nm was calculated to be smaller than 2%. Therefore the data for* $t = 36$ *nm is not plotted.*

saturation magnetization is almost[13] identical for a film thickness of 15 nm and 30 nm. A saturation field of 2.1 T for an iron layer is consistent with the previous result of the free-energy calculations. Particularly with regard to the field of the objective lens at an acceleration voltage of 200 kV, one has to assume an incomplete magnetic saturation of an iron film in a TEM and thus a reduced dichroic signal.

[13]The aberrations of less than 0.5% can be explained by internal round off in the simulations.

t(Fe)	saturation field
15 nm	2.12 T
30 nm	2.13 T

Table 4.5.: *Saturation field $\mu_0 H$, depending on the thickness of the unbounded iron layer according to LLG simulations.*

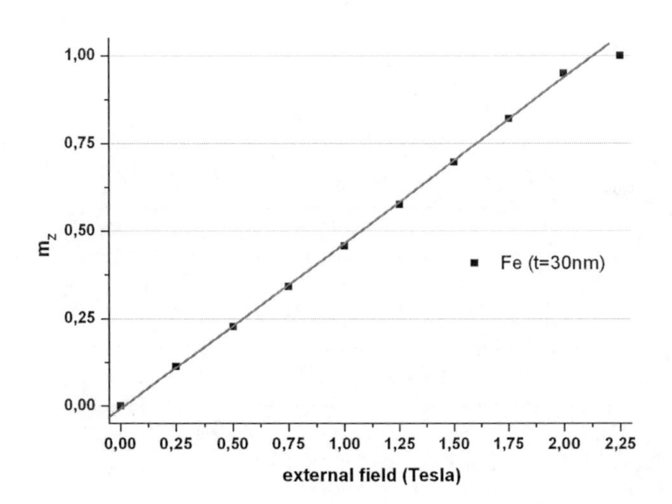

Fig. 4.14.: *Normalized perpendicular component $m_Z = \left|\frac{M_Z}{M_S}\right|$ of the magnetization in a 30 nm iron layer with periodic boundary conditions against the external field $\mu_0 H_Z$, calculated with LLG. The linear fit (including only data points before saturation) shows a line through origin with a gradient of $0.47\,\frac{1}{T}$.*
The difference in m_Z to an iron layer with $t = 15$ nm was calculated to be smaller than 0.5%. That's why the data for $t = 15$ nm is not plotted.

4.5. Summary of chapter "Magnetic considerations"

From calculations on the free energy of a ferromagnetic specimen and additional micromagnetic simulations using the LLG code, the degree of magnetic saturation parallel to the field of the objective lens of the TEM was determined. Although the free energy Ansatz contains several simplifications, the results agree very well with the output of the LLG simulations.
For nickel specimens, no deviation from a complete perpendicular saturation is expected

- independent of specimen thickness and acceleration voltage (in the range that is interesting for EMCD). In contrast to nickel, an iron specimen will not be saturated[14] along the external field of the objective lens. Using the simulation data shown in figure 4.14 and the measured magnetic field at the specimen's location (see section 4.1), the degree of saturation of a 30 nm iron layer can be estimated for different acceleration voltages in each case. The results are displayed in table 4.6.

acceleration voltage	m_Z
100 kV	63%
200 kV	82%
300 kV	93%

Table 4.6.: *With* $m_Z = \left|\frac{M_Z}{M_S}\right|$, *the percentage of the perpendicular component* M_Z *of the magnetization relative to the saturation magnetization* M_S *in a boundless iron film is shown for different acceleration voltages (and thus different fields of the objective lens, according to table 4.1). Even at* 300 *kV, iron layers are not completely saturated parallel to the field of the objective lens.*

Referring to the simulations, a cobalt specimen would be completely saturated in the field of the objective lens. This statement is contradictory to experimental research (remember figure 4.4). This discrepancy can be explained by the real appearance of the cobalt specimen's: Neither ideally flat, nor boundless - but supposable containing pinning centers. Also the film thickness was unknown in the experiment mentioned.

It is noticeable that the calculations on the free energy of a ferromagnetic specimen lead in all cases to the same results for the saturation magnetization as the micromagnetic simulations do (at least as long as no magnetic domains appear in the film).

[14] As the divergency from a complete magnetic saturation is only small, magnetic domains will hardly be observable in TEM imaging mode.

5. Specimen preparation and characterization for EMCD experiments

One of the main tasks of the group of the University of Regensburg within the ChiralTEM project is the preparation and characterization of specimens for EMCD measurements. In this chapter, the requirements for EMCD specimens are defined and different possibilities for specimen preparation are described and assessed.

5.1. Requirements for EMCD specimens

Every specimen for Transmission Electron Microscopy has to accomplish strict requirements concerning lateral dimensions and stability [Rei97]. On the one hand, as a guiding value, a specimen should be thinner than about 200nm[1] to get enough electrons penetrating the specimen for the illumination of the CCD camera.

On the other hand, the specimen has to fit the 3 mm circular entry of a TEM specimen holder (either by cutting it in an adequate shape or by mounting it on an annulus). And the mechanical stability should allow several cycles of inserting and rejecting the specimen, because the specimens have to be detached and stored in an exsiccator after each session at the TEM.

For EMCD measurements, there is a couple of further requirements for an ideal specimen. The following requirements refer to an ideal, well defined specimen for a maximum dichroic signal as demanded by the ChiralTEM project for an experimental verification of EMCD. Today (2008), with an improved spatial resolution and better signal to noise ratio, the requirements on a specimen are not as strictly as at the beginning of the project in 2004.

- **Ferromagnetic specimen**

- **Perpendicular saturated magnetization**
  As the dichroic effect reaches a maximum if the magnetization of the specimen is parallel to the electron beam, it has to be ensured that the specimen's magnetization is completely saturated parallel to the electron beam. A tilt between the magnetization direction and the electron beam or the occurrence of magnetic domains or areas with an in-plane magnetization would reduce the dichroic signal. With regards to specimen preparation, the most important way of influencing

[1]depending on the atomic mass of the element [Rei97]

the specimen's magnetization is a well considered choice of the used element (see chapter 4 for further information on the specimen magnetization).

- **Single crystal specimens**
  As, up to now, for all experimental setups for dichroic measurements in a TEM the specimen has to act also as a beam-splitter, specimens with a size of the crystallites larger than the illuminated area of the specimen are required. Within the ChiralTEM project, the required crystal size was thus determined to be larger than 1 μm. Although expensive and laborious to prepare, single crystal specimens would be the best choice. Polycrystalline specimens can be used if the crystallites are large enough and if, along the path of the electron through the specimen, not various crystals are being illuminated.

- **Optimum specimen thickness**
  In addition to that, it is not sufficient that the specimen is merely transparent for electrons. According to theory (see chapter 2.4 for further details), at a predefined acceleration voltage of the electrons, the size of the dichroic effect strongly depends on the thickness of the specimen. For an electron energy of 200 keV, the maximum dichroic effect is expected at 8 nm of nickel, 18 nm of cobalt and 20 nm of iron.

- **Flat and even specimens**
  For the same reason, a flat surface and an even shape of the specimen is required to prevent averaging the dichroic effect over different thicknesses. A wedge-shape, as it is usually generated by plane-view preparation, is only feasible if the illuminated area remains quasi even[2].

- **No substrates**
  As the operator of a dichroic measurement has to deal with a weak electron beam after the energy filter and thus either a bad signal to noise ratio or drift problems[3], any substrate layer below the magnetic specimen would complicate a dichroic measurement even more.

5.2. Specimen preparation by vacuum deposition

5.2.1. Self supporting Ni and Co specimens

One common option for TEM specimen preparation is the vacuum deposition. Therefore, within the scope of a diploma thesis, the influence of different parameters on the grain size was investigated (see [Hus06] for details).

For a fixed combination of deposit and substrate, the crucial factor for the grain size is the surface mobility of deposit atoms on the substrate [Hea70], [Hol70]. This mobility can be influenced by varying the substrate temperature, the deposition rate or the film

[2]The thickness difference has to be small compared to the illuminated area.

[3]Due to a drift, either of the energy loss scale or of the specimen holder, the exposure time should not exceed 60 s.

thickness. Annealing after the growth process can also increase the grain size. The qualitative dependences are shown in figure 5.1.

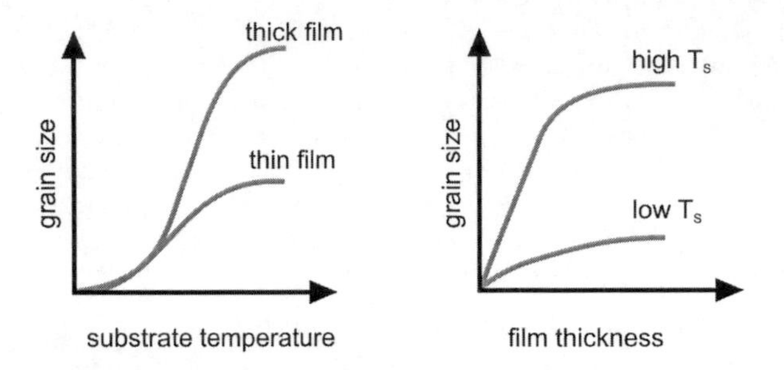

Fig. 5.1.: *Qualitative dependence of the grain size of*
- left side: the substrate temperature for thick and thin films
- right side: the deposition rate for high and low substrate temperature T_S
freely adapted from [Cho69]

According to [Cho69], the grain size increases with a raise of the substrate temperature and with a raise of the film thickness (for thin films the dependence on the thickness is almost linear). At small deposition rates, an increase of these values also produces larger grains. Numbers cannot be given for anyone of these parameters as other parameters like smoothness of the substrate surface, contamination, residual gases and the combination of evaporant and substrate have an immense influence on the results. Thus, an experimental series was done in order to find the optimum adjustment of these parameters and the optimum combination of deposit and substrate for large grains in self-supporting nickel and cobalt films.

The available high vacuum coating device (see [Ste05] or [Hus06] for construction details) is equipped with 4 water cooled beryllium-nitrite melting crucibles filled with Ni, Co and Al as evaporants (the latter was used for the deposition of an oxidation protection layer (approx. 5nm) and the fourth crucible remained unused). With a turbomolecular pump, a residual pressure of 10^{-6} mbar was reached. The substrate temperature can be adjusted from room temperature up to $(340 \pm 10)°C$ and deposition rates between $0.2 \frac{nm}{min}$ and $4.5 \frac{nm}{min}$ are possible.

For a substrate, either a freshly cleaved NaCl crystal was used, or a lacquer[4] coated glass plate. The latter can only be used up to a substrate temperature of 270°C, otherwise the lacquer would evaporate.

Equipped with a quartz crystal thickness monitor in the substrate plane with known area A, mass m_Q and initial resonant frequency ν_Q, the film thickness t_F of a material

[4] 2 g Polyvinylpyrrolidon in 60 ml ethanol

with the specific density ρ can be calculated from the change of the resonant frequency $\Delta\nu$ by: [Fre87]

$$t_f = \frac{m_Q}{\rho \cdot A} \frac{\Delta\nu}{\nu_Q} \tag{5.1}$$

with $\rho_{\text{Ni}} = 8.91\,\frac{\text{g}}{\text{cm}^3}$ and $\rho_{\text{Co}} = 8.89\,\frac{\text{g}}{\text{cm}^3}$

The ferromagnetic properties of the grown films can be demonstrated by Lorentz microscopy (see chapter 3.1 for details). All grown films clearly show magnetic domains, as it can be seen from the example of a 40 nm nickel film (compare figure 5.2).

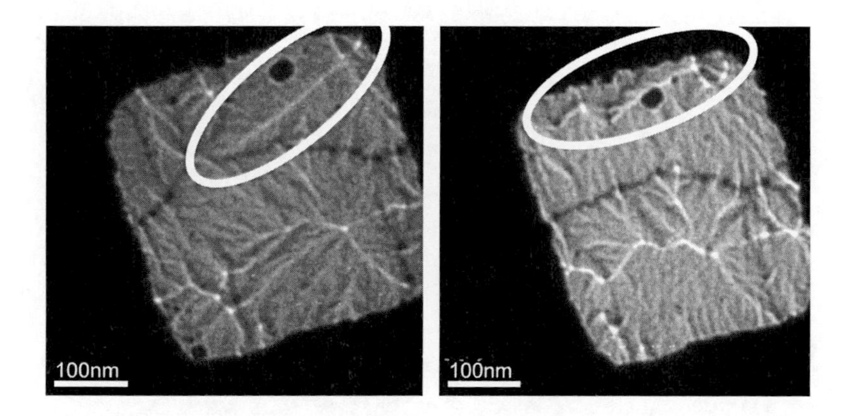

Fig. 5.2.: *Lorentz micrographs at a defocus of* 15 μm *show magnetic domains in a nickel film (grown on NaCl substrate) in copper grid [Hus06]. Between the two images, the tilt angle was changed by about* 5° *in the field of the weakly excited objective lens. The movement of the domains (and thus the verification of the ferromagnetic property of the grown film) can be observed in the white marked areas. Due to the tilt, also the dirt particle in the upper part changes its shape and its apparent position.*

The distribution of the different grain sizes was counted from different micrographs, taken at a magnification of 89000 and with an objective aperture set to make the single grains distinguishable (compare section 3.1 for the setup of a TEM). With (42 ± 4) nm for nickel and (76 ± 8) nm for cobalt films, the largest grains were achieved with the deposition parameters shown in 5.1. Whereas for both, nickel and cobalt, a substrate temperature of $340°C$ (and thus the technically possible maximum) is ideal[5], the best

[5]Even higher substrate temperatures would have been probably better but were not achievable with the available system.

deposition rate is determined to be 4.5 $\frac{\text{nm}}{\text{min}}$ for nickel and 2.3 $\frac{\text{nm}}{\text{min}}$ for cobalt films.

Even at a film thickness of 40 nm for nickel and 58 nm for cobalt, which were the thickest layers in the experimental series (and already too thick for EMCD), the achievable maximum grain sizes are at least one magnitude too small, regarding the claim for a grain size of 1 μm for the ChiralTEM project.

	film thickn.	substr. temp.	depos. rate	av. grain	max. grain
Ni	(40 ± 8) nm	$(340 \pm 20)°C$	(4.5 ± 0.3) nm/min	(12.7 ± 1.3) nm	(42 ± 4) nm
Co	(58 ± 12) nm	$(340 \pm 20)°C$	(2.3 ± 0.1) nm/min	(27.8 ± 2.8) nm	(76 ± 8) nm

Table 5.1.: *For nickel and cobalt films, the largest grains (average grain size and maximum grain size) in the experimental series are being achieved with the parameters substrate temperature and deposition rate. The associated distribution of the grain sizes can be seen in figure 5.3 for nickel and figure 5.4 for cobalt*

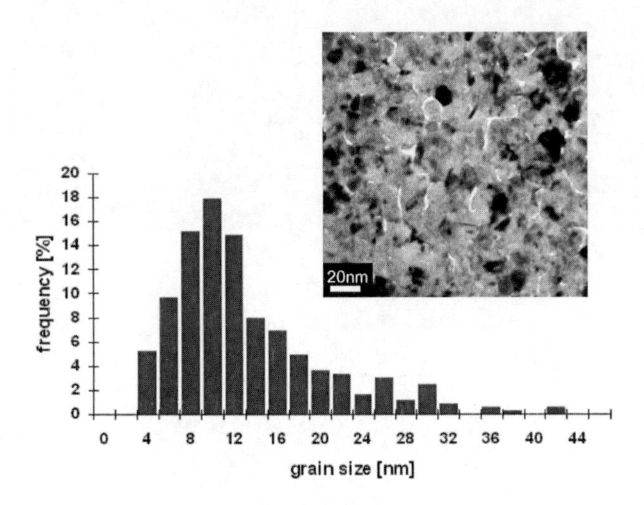

Fig. 5.3.: *Distribution of the grain sizes in a nickel film, grown with the parameters of table 5.1. In the corresponding electron micrograph, an objective aperture was set to visualize the different grains. The average grain size is* (12.7 ± 1.3) *nm and the maximum grain size is* (42 ± 4) *nm [Hus06].*

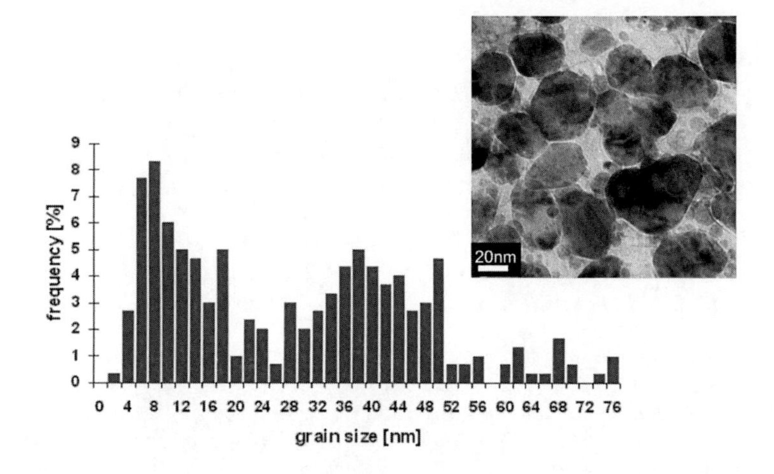

Fig. 5.4.: *Distribution of the grain sizes in a cobalt film, grown with the parameters of table 5.1. In the corresponding electron micrograph, an objective aperture was set to visualize the different grains. The average grain size is* (27.8 ± 2.8) *nm and the maximum grain size is* (76 ± 8) *nm. The first peak in the distribution is caused by nucleation, the second peak is caused by the formation of clusters [Hus06].*

5.2.2. Epitaxial Fe specimens on a GaAs diaphragm

As the preparation of self supporting specimens by vacuum deposition did not lead to satisfying results, an other option was the vacuum deposition of Fe on an electron transparent GaAs diaphragm. The commercially available diaphragms usually used for TEM specimens have a thickness of $(30-35)$ nm of amorphous Si_3N_4 [Uhl04] and are therefore not applicable for the growth of epitaxial layers on their surface.

Alternatively, at the University of Regensburg one[6] grew epitaxial iron layers on a gallium arsenide (GaAs) layer system that can be etched chemically to a diaphragm after the growth of the iron layer as depicted in figure 5.5 [Mei06].

The result of the laborious preparation can be seen in figure 5.6. The creation of an electron transparent membrane below the epitaxial iron layer works in principle. In any case, the surface of the GaAs membrane is very rough due to etching residua. As the specimen is very damageable by any vibrations and the residua are located in the etching crater, they can not be removed without destroying the diaphragm. Thus, this preparation technique is not applicative for EMCD experiments.

[6]The specimen preparation was done by Roland Meier.

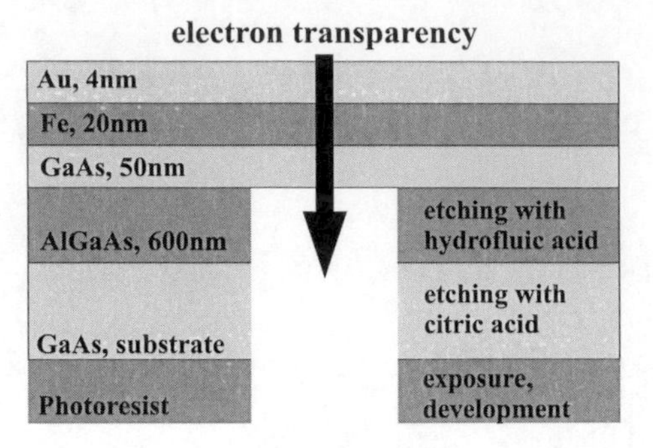

Fig. 5.5.: *Schematic view of the preparation of an iron layer on a GaAs diaphragm. The layer system is produced by molecular beam epitaxy. On the substrate layer of GaAs, an etching mask is created by coating with photoresist, exposing and developing. The areas without photoresist layers are corroded by citric acid ($C_6H_8O_7$). As the GaAlAs layer is not vulnerable by citric acid, the etching process stops automatically. Using hydrofluic acid (HF), which etches only AlGaAs but stops at GaAs, the preparation process continues until only a self-supporting* 50 *nm layer of GaAs remains below the epitaxial iron layer with the gold capping for oxidation protection.*

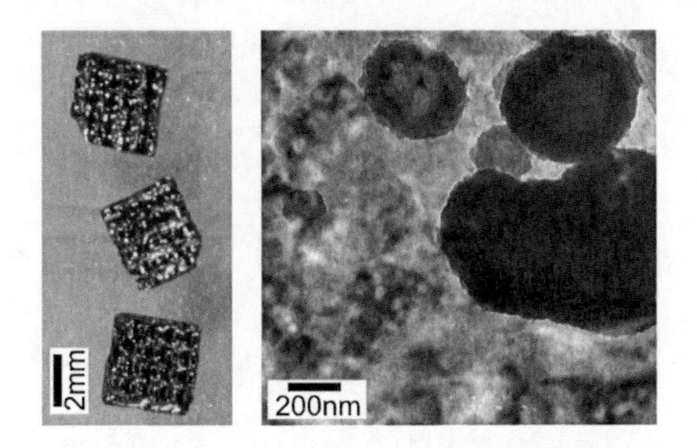

Fig. 5.6.: ***Left side:*** *Image of three specimens with 16 etching craters on each. The diaphragms are extremely damageable by every kind of mechanical stress.*
Right side: *Electron micrograph of an iron layer on GaAs diaphragm. Although the overall specimen thickness is suitable for TEM investigations, the rough undersurface of the GaAs diaphragm makes these specimens unfeasible for EMCD.*

5.3. Specimen preparation by bulk adaptation

In contrast to the specimen preparation by vacuum deposition stands the possibility of thinning a massive piece of the desired element (ideally available as single crystal) to an electron transparent film. For this approach, different techniques such as mechanical polishing, ion etching (including the focused ion beam (FIB) technique) and electrochemical etching are being investigated. The possibility of laser induced chemical etching was discussed within the ChiralTEM project, but it was abolished due to inadequate experimental equipment.

5.3.1. Dimpling and ion etching

A common specimen preparation technique for transmission electron microscopy is the treatment of a thin film by ion etching [Kai00]. In this case, single crystals of cobalt (the origin thickness of the available crystals is 1 mm) are being cut into pieces with an edge length of 2 mm and afterwards they are polished with a diamond grinding disc down to a remaining thickness of 100 μm. In the resulting cobalt foil, two opposing dimples are grinded, using a Gatan dimple grinder. The advantage of this step is the conservation of the mechanical stability of the whole specimen while reducing the thickness in a small area down to approximately 30 μm. The last preparation step is the two-phase ion etching, using a *Balzers Baltec* (with xenon as etchant) or alternatively a *Gatan Precision Ion Polishing system (PIPS)* (with argon as etching gas). In the first phase, a small hole is etched in the dimples on the foil, using an acceleration voltage of 3 kV - 4 kV (the etching time is about 20 h, depending on the starting thickness in the center of the dimple). In a second phase, the area around the hole is polished with a reduced acceleration voltage of 1.5 kV (polishing time approx. 2 h) to receive a clean, smooth surface without grooves. All preparation steps are outlined in figure 5.7. In the investigated specimens, no significant difference between the two gas sources (Xenon or Argon) was apparent.

Even for cobalt, which has according to the calculations (section 2.4) the lowest requirements concerning the specimen thickness, the resulting specimens are - except the area directly around the hole - with more than 50 nm too thick for dichroic experiments. In addition to this, the former single crystal is modified to a polycrystalline film. Especially the area directly at the hole is affected that way and thus any dichroic measurement - which requires two subsequent measurements with exposure times of some 10 seconds each on one and the same crystal (see chapter 6) - would only be possible if the drift of the specimen was smaller than the crystal size. As it is actually even difficult to find a position on the ion etched cobalt specimen where only one crystal is illuminated by the incident electron beam in diffraction mode, these specimen are classified as not suitable for EMCD. A further reduction of the acceleration voltage and with it the possibility for a reduction of the crystal modifications is not possible as the ion etching time is already within some 10 hours and the capacity of the available ion mills allows no further extension of the preparation time.
Nevertheless, the resulting cobalt specimen shows a magnetic maze domain pattern as

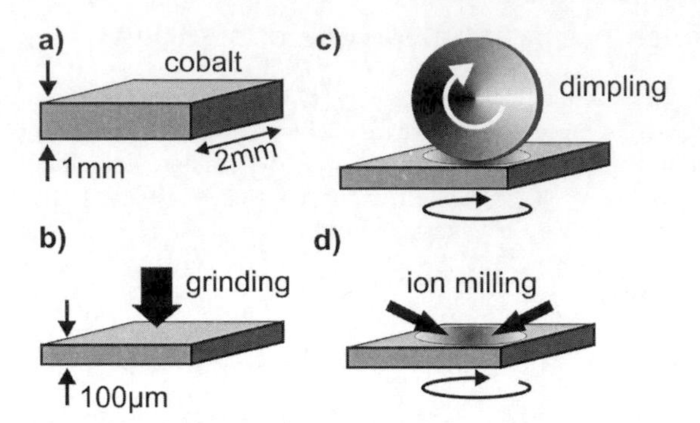

Fig. 5.7.: *a) The single crystal is cut to pieces with a size of 2mm to 2mm.*
b) The cuboid is grinded to a film of approx. 100 μm
c) The grinding wheel makes a dimple into the film. Remaining thickness about 30 μm.
d) Finally, a hole is shot in the dimple using an ion mill with either Argon or Xenon as etching gas.

expected according to the micromagnetic simulations (chapter 4). A direct lineup of simulation and Lorentz micrograph can be seen in figure 5.8.

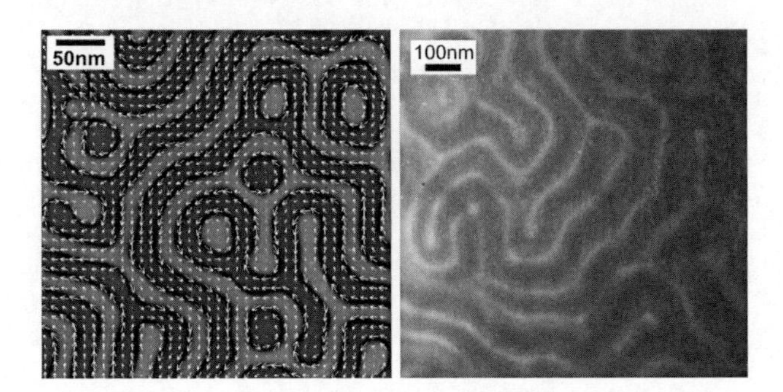

Fig. 5.8.: *Comparison of maze domains in a micromagnetic simulation (chapter 4) and a Lorentz micrograph. The different width of the stripes can be ascribed to a different layer thickness (48 nm in the simulation and unknown but expected thicker in the micrograph). The similar appearance validates on the one hand the simulations. On the other hand the appearance of magnetic domains is a verification of the ferromagnetic property of the produced specimen.*

At the beginning of the ChiralTEM project EMCD in Lorentz mode[7] was considered. Therefore, also yttrium-bismut-iron garnet (YBiFeG) specimens have been investigated (the preparation is done the same way as described for the cobalt specimens in the current chapter by polishing, dimpling and ion etching). YBiFeG has an uniaxial anisotropy like cobalt and has thus the nature to develop maze domains with areas of perpendicular magnetization without the presence of an external field and generates magnetic bubbles when applying an external field perpendicular to the specimen plane (see chapter 4.4). But unlike cobalt with its saturation field of more than 1 T, the saturation magnetization of the investigated YBiFeG specimen is only 0.2 T. Hence, the domains in a YBiFeG specimen can easily be controlled[8] by the field of the objective lens (which is not used for imaging in Lorentz mode, see chapter 3.1). The change of the domain structure in an external field can be seen in figure 5.9. The results of methodical investigations on the influence of an external field on the width of parallel and antiparallel maze domains and the generation of magnetic bubbles can be found in graph 5.10. The investigated YBiFeG crystal has the chemical composition $(Y_xBi_{3-x})Fe_2Fe_3O_{12}$ with unknown $x = 0...3$ [Doe06].

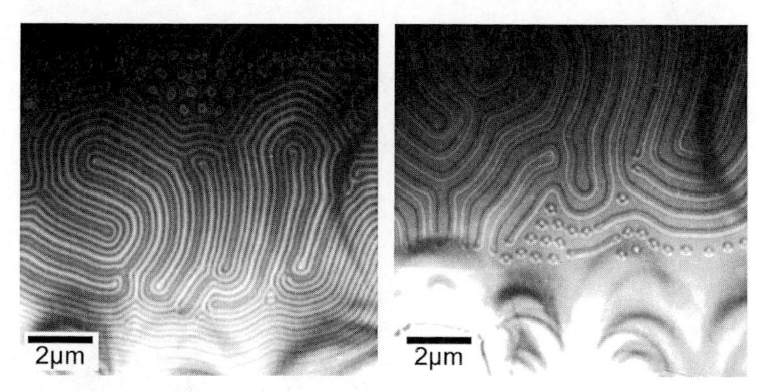

Fig. 5.9.: *Lorentz micrographs of maze and bubble domains in a YBiFeG specimen, without external field (left side) and with an external field of approx.* 70 *mT (right side). The bubbles in the upper part of the left image are remanent leftovers of a precedent saturation of the specimen. From the left to the right image, a growth of areas parallel to the external field at cost of the areas antiparallel is visible. In the middle of the right image, maze domains decompose into bubble domains.*

[7]This option was skipped early in the project as it is one step beyond the claimed targets of the project - the experimental verification of the dichroic effect. A more extensive reflection on EMCD in Lorentz mode can be found in the outlook, chapter 8

[8]enlarging areas parallel to the external field, creating magnetic bubbles or shifting these bubbles by tilting the specimen in the field of the slightly excited objective lens

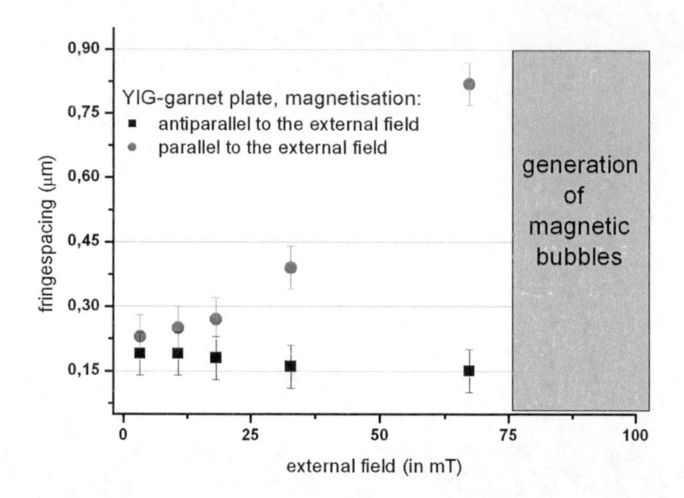

Fig. 5.10.: *This graph shows a quantitative analysis of the variation of domain widths in an increasing external field $\mu_0 H_{ext}$ in a YBiFeG specimen, as displayed in the micrographs in figure 5.9. At a (out of plane) field of approx. 75 mT (depending on the specimen thickness), the maze domains decompose into bubble domains which remain up to an external field of approx. 200 mT.*

As the investigations on EMCD in Lorentz mode have been skipped in an early phase of the ChiralTEM project for the benefit of the objective lens reversal technique (described in detail in chapter 7), no dichroic measurements were performed with YIG specimens.

5.3.2. FIB preparation of Ni and Co

In the precedent section it is explained that the modification of the crystal structure during the ion etching process is too severe for dichroic experiments. As the focused ion beam technique (FIB) (see [Lan01a] and [Lan01b] for detailed information) also exposes the specimen to a high energy ion beam, the entitlement for these further experiments can only be explained when comparing the different ways the ions mill the specimen using FIB etching or conventional ion etching:

Whereas the ion beam in conventional ion mills hits the specimen on a large area (especially also in the interesting area around the emerging hole, see figure 5.7d) and can thus cause a crystal modification through ion implantation and heating all over this area, the ion beam of a FIB is able to cut only selected areas off the specimen. In this case, a nickel or cobalt crystal (2 mm by 2 mm) is grinded down to a thickness of about 80 μm and afterwards polished to a wedge shape with one front surface having a remaining thickness of approx. only 10 μm (compare figure 5.11). This surface is coated with a titanium mask with the shape of the desired final slice. In contrast to the titanium capped area, the unprotected area of the specimen is removed by the ion beam. The remaining, self-supporting windows have a width of about 10 μm to 20 μm (depending on the titanium mask) and a remaining thickness of about 30 nm to 50 nm at the outer boundary.

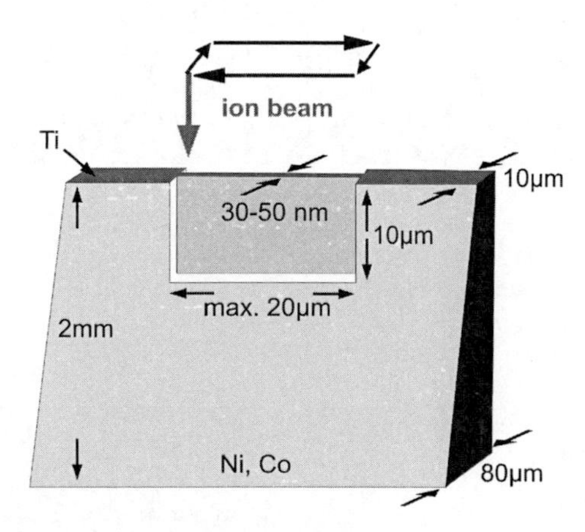

Fig. 5.11.: *Principle of the applied FIB specimen preparation technique: a wedge-shape polished bulk single crystal of nickel or cobalt is coated on the thin border with a titanium mask with the shape of the desired window. Afterwards, the window is cut by a controlled focused ion beam.*

Using FIB, the material, that will remain after the etching process, is not directly hit by the ion beam and the heat dissipation is better as the surrounding bulk is thicker (compared with conventional ion milling). Even so, FIB cut cobalt layers show a serious modification of the crystal structure as can be seen in figure 5.12. The crystal modification is much smaller in a nickel specimen (compare figure 5.13), but in nickel, the achievable thickness of approx. 50 nm at the edge of the FIB cut window is too large for dichroic measurements.

Altogether three specimens have been prepared by FIB[9], two made of cobalt and one made of nickel with several windows in each but none of them fulfilled the requirements of the ChiralTEM project. Due to the extensive efforts at rather low chances, no further FIB specimens have been prepared for dichroic experiments.

[9]The FIB preparation was done by Dr. R. Kroeger, York (UK), formerly Bremen

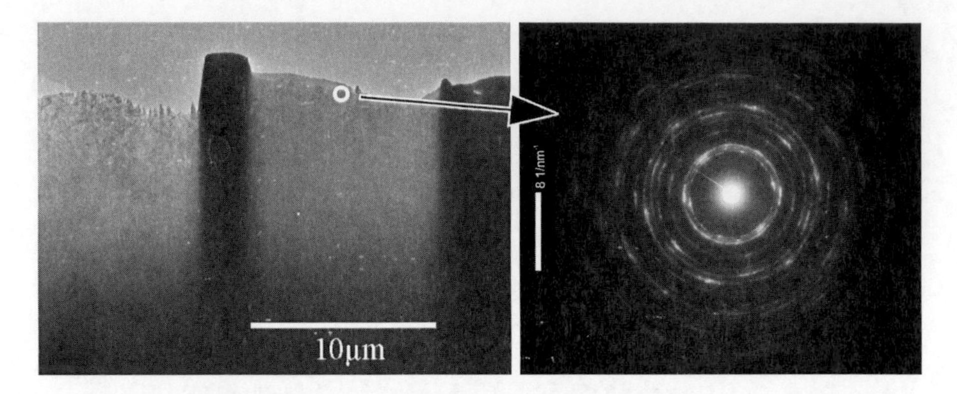

Fig. 5.12.: *Electon micrograph (left side) and diffraction pattern (right side) of a FIB cut cobalt film. Whereas the layer thickness at the border is with about* 30 *nm in a suitable range for EMCD, the modification of the crystal structure makes any dichroic measurement impossible.*

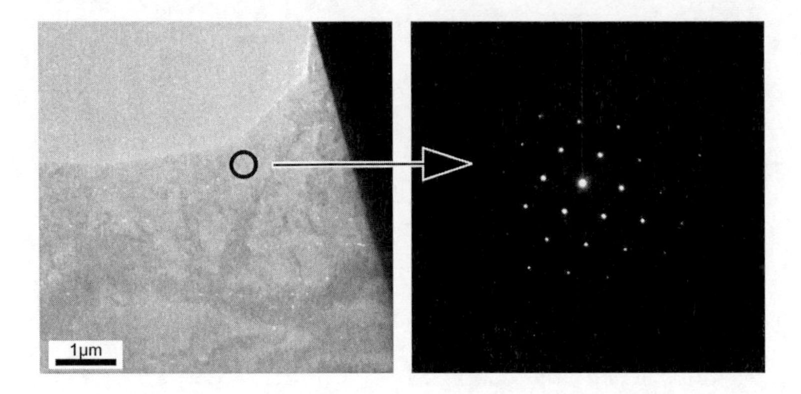

Fig. 5.13.: *Electon micrograph (left side) and diffraction pattern (right side) of a FIB cut nickel film. In contrast to the cobalt specimen (see figure 5.12), the crystal modification is negligible. But the achievable layer thickness is approximately* 50 *nm and thus no dichroic signal is expected according the calculations, section 2.4.*

5.3.3. Electrochemical etching of Ni

An other approach to the bulk preparation is electrochemical etching. For the specific requirements (thin and flat monocrystalline areas, see section 5.1), an etching setup with optical etching control is required. Using such a system, the etching current is triggered by an infrared detector which stops the etching process as soon as the light of an infrared light source penetrates the developing hole in the specimen (compare figure 5.14). As this system is not available in Regensburg but at the TU Vienna (member of the ChiralTEM collaboration, see chapter 1), the preparation by electrochemical etching was performed in Vienna.

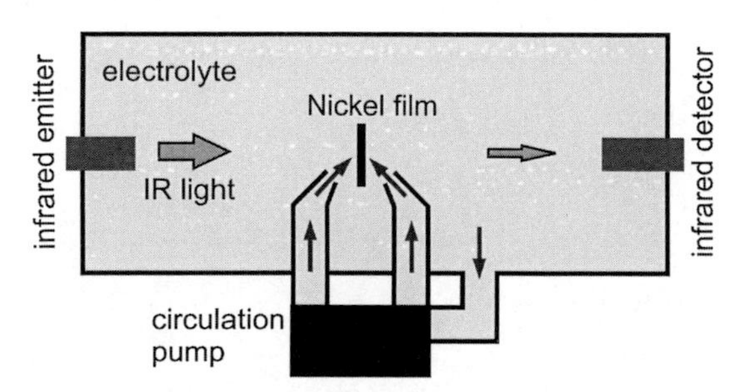

Fig. 5.14.: *Principle of electrochemical etching. The nickel film is washed by the circulating electrolyte. An infrared detector stops the etching process as soon as the film starts getting transparent.*

The process is described for the preparation of a nickel specimen as this material showed the best results in Vienna. Starting with a mechanical thinning of a nickel single crystal to a nickel foil as already described in section 5.3.1, the foil is immersed into the electrolyte bath. As electrolyte, the following non-corrosive composition is being used [Rub07c]:

- 14 g of *Magnesium Perchlorate*, $Mg(ClO_4)_2$ as oxidant
- 625 ml of *Methanol*, CH_3OH as solvent
- 125 ml of *Cellosolve (ethylene glycol monoethyl ether)*, $C_2H_5OCH_2CH_2OH$ for a control of the freezing point and
- 6.6 g of *Lithium Chloride*, LiCl

At a voltage of 80 V and a temperature of 243 K ($-30°C$), the perforation occurs within some 10 seconds (depending on the initial thickness of the nickel film). After the etching is stopped by the infrared detector, the specimen is cleaned by an ion etching system (Gatan Duomill) to remove residua of the chemical etching process. Experimental details and different recipes can be found in [Tho77] or [Schi81].

Using this preparation technique, the crystal structure remains extensively unaffected - the size of the single crystals is in the range of several 10 μm and thus adequate for dichroic experiments. Also the accessible specimen thickness is with about 10 nm in the direct surrounding of a hole ideal for EMCD. Disadvantages are on the one hand - comparable to the bulk preparation by ion milling, section 5.3.1 - the wedge shape of the area around a hole and thus no homogeneous specimen thickness. On the other hand, at the thin areas, the specimen gets bent. Therefore, the advantage of the large crystals is reduced as after a shift of the specimen (for example due to a drift of the specimen holder) the tilt angles have to be corrected to reach the same orientation of the crystal. An image series of the nickel specimen used for the dichroic measurement in chapter 6 can be seen in figure 5.15.

5.4. Summary of chapter "specimen preparation"

5.4.1. Preparation techniques

Finally, a summary of all the different possibilities for specimen preparation can be given in terms of a table 5.2. The aspects crystallinity (achievable grain size), thickness controllability, the disturbing influence of a substrate (diaphragm) and the possibility to produce flat layers are evaluated.

As can be seen in table 5.2, the vacuum deposition of self supporting layers is not suitable because the crystallites are getting too small for the adjustment of a proper two beam case in scattering mode.

Also the vacuum deposition of Fe on GaAs diaphragms is not feasible, because the diaphragm remaining after the last etching process is both too thick and not even enough (although the Fe layer on the surface is sufficiently flat and even).

The bulk preparation by ion etching causes severe modifications of the crystal structure in the areas with an adequate thickness. This disadvantage can be avoided when using electrochemical etching instead of ion etching.

Not finally investigated is the possibility of cutting specimens using a dual beam focused ion beam engine. The first trials showed either a strong modification of the crystal structure (using cobalt crystals) or they were too thick for EMCD (using nickel crystals). Maybe a further experimental series can produce suitable specimen. Within this dissertation, no further FIB cuts were possible as the FIB engine in Bremen used for the preparation is not directly accessible and commercial FIB cuts are very expensive.

Summing up all performed experiments, electrochemical etching from a bulk single crystal is the best preparation technique for EMCD specimens.

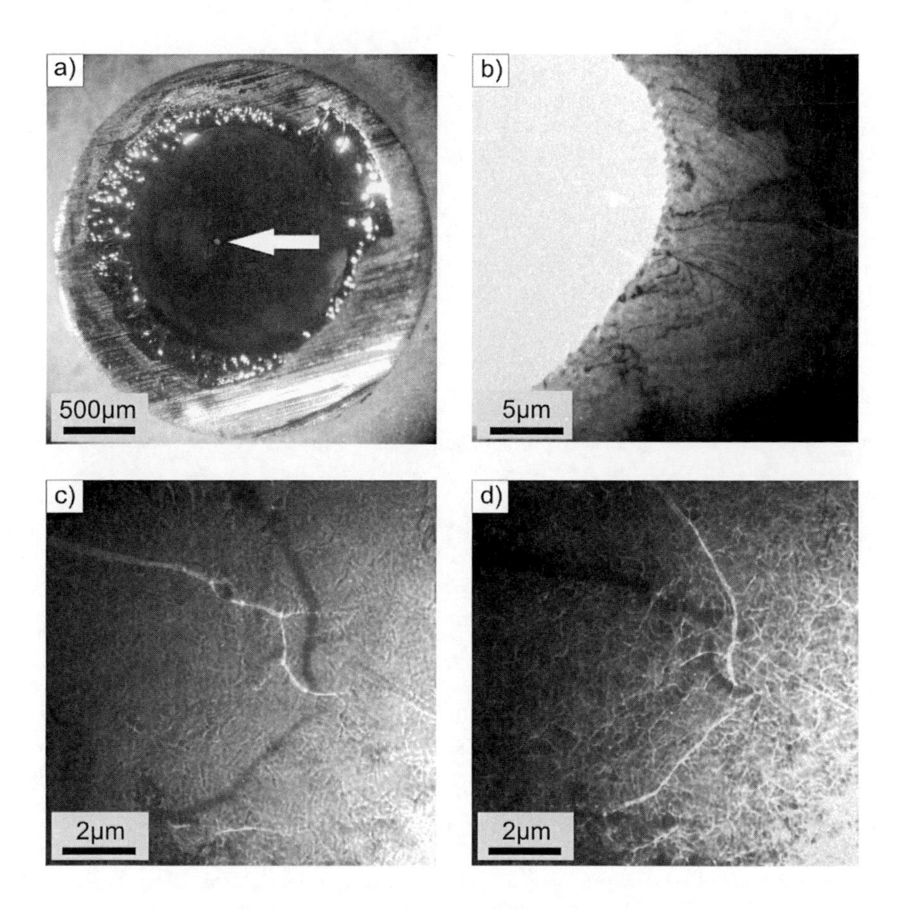

Fig. 5.15.: *Images of a nickel specimen produced by electrochemical etching:*
a) optical micrograph of the nickel film mounted on an Al annulus. The hole is marked by the white arrow.
b) TEM micrograph of the area around the hole.
c) Lorentz micrograph near the hole in underfocus - walls between magnetic domains appear as bright and dark lines.
d) Lorentz micrograph near the hole in overfocus - the contrast of the domain walls is reversed.
c+d) The filigree structure in the two Lorentz micrographs is generated by grooves from the electrochemical etching process and has no magnetic origin.

	crystallinity	thickness	substrate	flatness
self supporting deposition, *chapter 5.2.1*	-	+	+	0
deposition on diaphragm, *section 5.2.2*	+	+	-	0
ion etching, *section 5.3.1*	-	0	+	0
FIB preparation, *section 5.3.2*	-/0[1]	0/-[1]	+	+
electrochemical etching, *section 5.3.3*	+	0	+	0

Table 5.2.: *Survey of the useability of different techniques of specimen preparation for EMCD specimens, "+" means advantageous, "0" means adequate and "-" means unemployable concerning the particular aspect. The required aspects are described in chapter 5.1.*
[1] *Whereas for cobalt, the thickness is adequate but the crystallinity is insufficient, for nickel, the grains would be large enough but the resulting layer is too thick.*
According to this table, electrochemical etching produces the best specimens for EMCD

5.4.2. Magnetic materials

Also the suitability for dichroic experiments of the different investigated materials can be shown in a table (table 5.3). Evaluated is the thickness that is required for a large dichroic signal according to the theoretic predictions (chapter 2.4), the possibility of the preparation of the particular materials to the prementioned thickness, the magnitude of the relative dichroic signal (according to theory) and the possibility for a complete magnetic saturation of the specimen according to the magnetic calculations of chapter 4.

Considering all these parameters the recommendation for the easiest way to accomplish a dichroic measurement in a TEM would be choosing a nickel specimen.

	possibility of prep.	requ. thickness	EMCD signal	magn. saturation
Fe	+	**0**[1]	+	**0**
Ni	+	**0**[1]	+	+
Co	+	+	**0**	**0**
YIG	**0**[2]	*unknown*[3]	*unknown*[3]	+

Table 5.3.: *Survey of the useability of different materials for EMCD experiments, separated in the possibility of specimen preparation, the ideal specimen thickness for EMCD according to theory 2.4, the strength of the predicted dichroic signal and the possibility for a complete magnetic saturation of the specimen by the field of the objective lens. "+" means advantageous, "**0**" means adequate and "-" means unemployable concerning the particular aspect.*

[1] *Within the due to theory required specimen thickness, a damage of the specimen is caused by the electron beam.*

[2] *The preparation of YIG is difficult as thin layers become very brittle.*

[3] *A dichroic effect in YIG was at the date of this work neither predicted by theory nor measured.*

6. EMCD measurement on a nickel specimen

According to the two precedent chapters, a nickel specimen prepared by electrochemical etching provides the best way to receive a first dichroic measurement in a TEM. In the present chapter 6, the execution of an EMCD measurement is described, a dichroic signal is shown and experimental problems are discussed.

6.1. Dichroic signal using the diffraction shift method

The applied specimen was prepared in Vienna according to the procedure explained in section 5.3.3. The thickness of the area dedicated to the measurement is determined with the help of a *thickness map*[1]. In the thickness map (compare figure 6.1), black means a thickness of zero (the hole) and the brighter a pixel appears, the thicker the specimen is at this point. In the direct neighborhood of the hole, the specimen thickness is about 15 nm and as the specimen forms almost a plateau, the variation of the thickness is negligible in a range of about 150 nm. According to theory (section 2.4), the optimum thickness for nickel would be 10 nm and therefore less than available. But one can assume that the thickness of the specimen will be locally reduced during the measurement due to the focused electron beam on the specimen (as will be shown in section 6.2).

Setting the energy filter to the $L_{3,2}$ edge of nickel, the number of electrons reaching the CCD detector is drastically reduced in comparison to the incident beam. In addition to this, in the present experimental setup, the spectrometer entrance aperture is set to a point between the Bragg spots of a diffraction pattern - a position that appears "black" in a diffractogram. To get a signal above the noise level anyhow, the beam emitting system has to be adjusted to the highest possible electron beam current.

For the measurement, the brightness of the electron beam is boosted by increasing the extraction voltage of the FEG (field emission gun) emitter to 4500 V and setting the parameters *gun lens*[2] to 3 and *spotsize*[3] to 1.

[1] In a thickness map, a zero loss peak with plasmon peak is recorded for each pixel of the map. The ratio between the area under the zero loss peak and the area under the plasmon peak specifies the thickness of the specimen at the location of this pixel.

[2] "The electrostatic gun lens converges the electron beam. If the gun lens is weakly excited, the cross-over occurs lower in the column and more electron current passes through the condenser aperture." [Tec08]

[3] Similar to the gun lens, a smaller excitation of the first condenser lens C1 (spotsize) results in a higher intensity on the specimen at cost of the spatial coherence [Tec08].

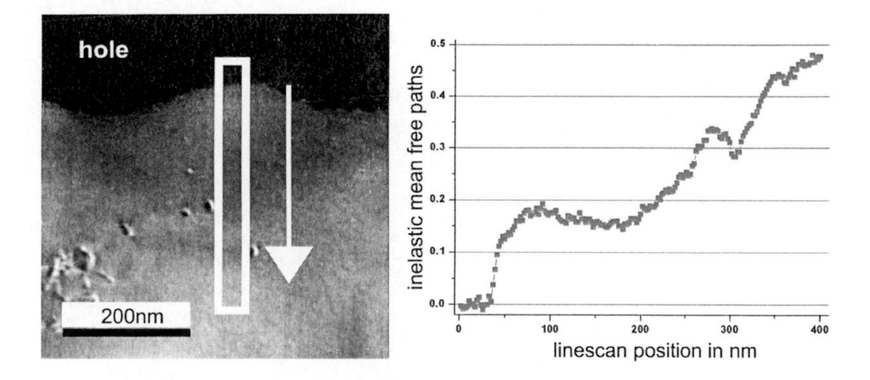

Fig. 6.1.: *Thickness map of the area selected for the dichroic measurement (left side) with corresponding linescan (right side). The brighter the specimen appears, the thicker it is. The linescan with a length of* 400 *nm shows the specimen thickness in multiples of the inelastic mean free path (the distance for the movement of an electron without lattice interaction). According to [Gat08], at an acceleration voltage of* 300 *kV, for an atomic number of* 28 *(nickel) and assuming a convergence semi-angle of* 2 *mrad and a collection semi-angle of* 40 *mrad, the inelastic mean free path is calculated to be* 94 *nm, with an unknown error in the implied parameters. Thus, the thickness of the plateau between the positions* 80 *nm and* 200 *nm in the linescan is in the range of* 13 *nm to* 17 *nm.*

Although with these settings[4] a FEG emission current of 63 μA was reached, the exposure time for each spectrum had to be set to 60 s to obtain a signal clearly above the noise level.

The specimen, mounted in the double-tilt specimen holder, is tilted to an angle (using the β-tilt) of approx 6.3° out of the crystal axis (001) and the two beam case $G = (0, 2, 0)$ is excited (using the α-tilt). The codensor aperture is set to 50 μm - a size that makes the diffraction spots large but not overlapping to gain intensity on the detector. The camera length is chosen to 380 mm. At this camera length, the ratio between the size of the Thales circle through the two Bragg spots and the 5 mm spectrometer entrance aperture is suitable to accurately locate the two positions A and B (according to figure 6.2)

The diffraction pattern is then shifted using *diffraction shift* to adjust the relative

[4]... and the emitter tip, attached at the time of the measurement ...

position of the spectrometer entrance aperture[5] being either position A or position B according to the definition in figure 6.2.

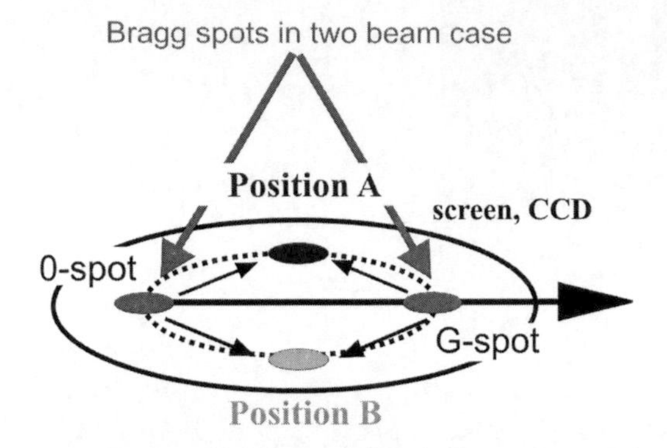

Fig. 6.2.: *Schematic view of the definition of the two measuring points (Position A and B) on a Thales circle through the Bragg spots in the diffraction pattern (two beam case). According to [Rub07a], the axis through the Bragg spots is defined positive from 0-spot to G-spot and the positions on the Thales circle are defined A (above) and B (below).*

Using this experimental setup, two spectra are recorded, one at position A and the other one at position B. Afterwards, a possible drift of the energy loss scale between the two measurements is corrected and the background is calculated and subtracted[6]. Then, the signal strength of the two spectra is adapted, using either an energy interval between the L_3 and L_2 edge or an energy interval above the L_2 edge (which must lead to the same result if the background correction was proper and successful). The dichroic signal is then the difference between the two corrected spectra and can be seen in figure 6.3. For the relative dichroic signal σ_{rel}, the definition

$$\sigma_{\text{rel}} = \left| \frac{\sigma^+ - \sigma^-}{\sigma^+ + \sigma^-} \right| \tag{6.1}$$

is used. With σ^+ and σ^- as signal of the particluar spectra, the measured dichroic effect can be calculated to be $\sigma_{\text{rel}}(L_3) = 4.8\%$ at the L_3-edge and $\sigma_{\text{rel}}(L_2) = 1.5\%$ at the L_2-edge and thus clearly above the noise level. The reason, why the measured relative dichroic signal is smaller than the theoretic prediction can be found in the following section 6.2. According to [Rub07a], the dicroic effect was also demonstrated on iron and cobalt specimens in Vienna within the ChiralTEM project. In this context, a reference measurement with a non-ferromagnetic titanium specimen showed no dichroic signal.

[5]For all EMCD measurements, the diameter of the spectrometer entrance aperture was set to the largest value - 5 mm.

[6]using the software Gatan Digital Micrograph [Gat08]

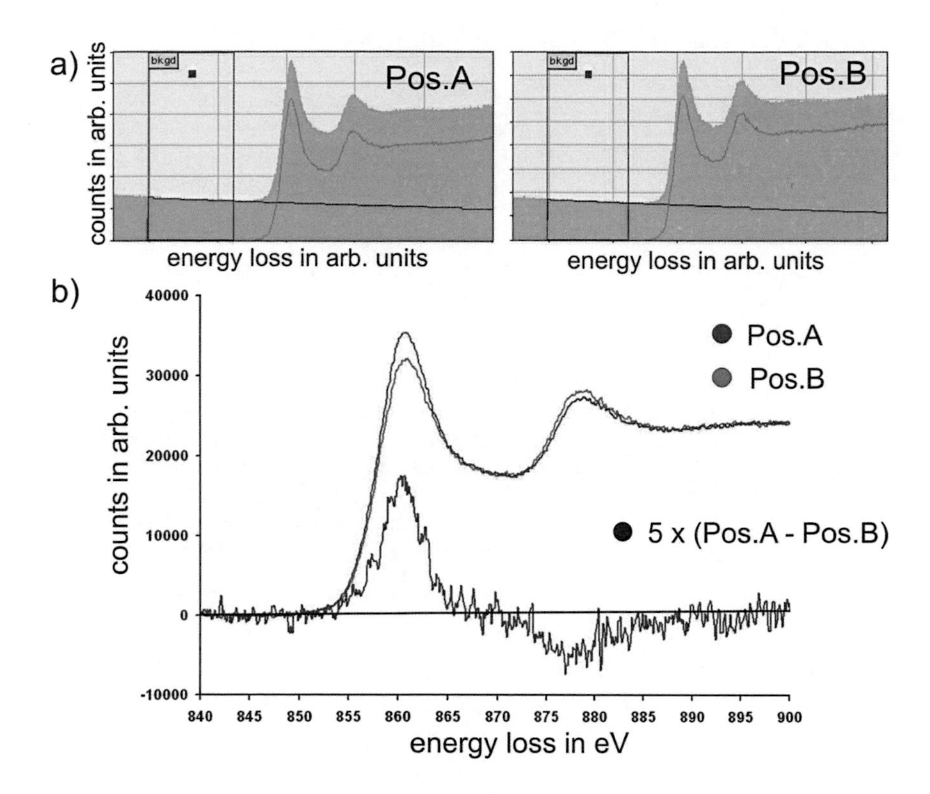

Fig. 6.3.: *a) The spectrometer raw data (turquoise), taken at the two opposite positions A and B on the Thales circle, are purified by subtracting the background (red), adapting the counts and calibrating the energy loss scale.*
b) After processing, the two spectra A and B show a dichroic difference signal (A-B) clearly above noise level. For a better visualization, the difference signal is magnified by a factor of 5.

6.2. Experimental stumbling blocks

Assuming an ideal specimen and performing a dichroic measurement on it (according to the experimental setup described in section 3.3), two critical experimental problems can occur: A drift of the energy-loss spectrum and a beam damage of the specimen. They may be partially avoided by a change or extension of the existing experimental setup. The complete elimination would require a completely different setup, attaining a higher signal thus less time for the recording of each spectrum, as it is proposed in chapter 8 and currently investigated at the University of Regensburg.

Spectrum drift One of the biggest (specimen-independent) problems is a drift of the energy loss spectrum during the recording of a spectrum. This happens due to a change of external magnetic fields. At the TEM lab in Regensburg, the situation is especially critical, as some labs around are working with magnetic fields (among a NMR (nuclear magnetic resonance) lab using fields up to 20 T in the direct neighbourhood). The shift of the energy scale during a measurement broadens the peaks in the spectrum and as the peaks at the L_3 and L_2 edges are not symmetrically, this broadening leads to a pseudo-dichroic difference between the two spectra[7].

This problem could be reduced by using a shorter illumination time, which would in turn require a brighter electron beam on the detector. A better solution would be the installation of an active magnetic field compensation, which measures the magnetic field at the specimen's location and compensates changes with a compensation field generated by three pairs of Helmholtz coils (see [Dun95] for details). The installation of such a compensation unit in Regensburg is already agreed upon, but it was not available at the time of this work.

Beam damage An other severe problem in measuring the dichroism is the damage of the specimens by the electron beam.

On the one hand, only few electrons appear in the area between the Bragg spots. Thus, the intensity of the incident electron beam has to be adjusted as bright as possible. On the other hand, the investigated (ideally self-supporting) specimen area has to be chosen very thin (the ideal specimen thickness for nickel is about 10 nm for an acceleration voltage of 300 kV), so the heat dissipation is rather bad. As a result, the electron beam cuts a hole into the specimen within a few minutes (either by fusing the specimen or by knock-on damage). An example for this damage can be seen in figure 6.4.

Thus, there is not much time for the orientation of the single crystal and the recording of two spectra with 60 s exposure time each and shifting the diffraction pattern in between without making any change on the illumination. As mentioned in the precedent section 6.1, this erosion can be used by selecting a position on the specimen with a thickness slightly larger than the desired ideal thickness. During the recording of the two spectra, the thickness will by chance run through the ideal thickness. Of course, a reproduction of a measurement at the same position on the specimen would be just as well impossible as a reliable measurement of the specimen thickness corresponding to a certain spectrum. This problem can be reduced by choosing a cobalt specimen with an ideal specimen thickness of about 20 nm but a considerably smaller relative dichroic signal or an iron specimen with an ideal thickness of about 22 nm and the additional need for a (signal

[7] This mistake can be easily identified as the integrated signal under the edges remains unchanged if no real dichroism is existent.

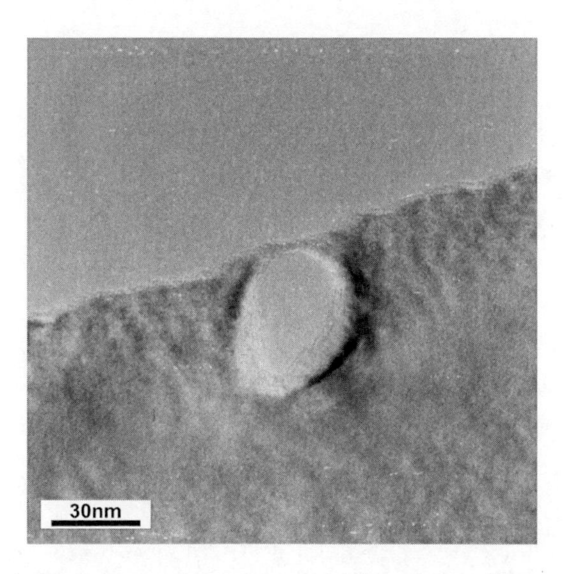

Fig. 6.4.: *Example for the beam damage in a nickel specimen. The specimen thickness at the border is approx.* 10 *nm and the small hole of about* 30 *nm by* 40 *nm was shot with a constricted electron beam (as it is used for dichroic measurements to gain signal) in 6 minutes. Nevertheless, the EMCD measurements taken on the same nickel specimen (near the area shown on this image) show a clear dichroic signal (compare figure 6.3).*

reducing) capping layer to prevent oxidation. Despite these problems, the measurement of the dichroic signal in the nickel specimen was easily reproducible by a small (approx. 200 nm) shift of the specimen to an area without beam damage after each measurement. A real solution of this problem could be achieved by a different experimental setup that uses no longer the specimen also as beam splitter. In this case, the dichroic signal would be to a large extent independent of the specimen thickness and thus thicker specimens with a better heat dissipation could be used. One approach for such a new setup is described in section 8.2.

After the successful EMCD measurement, the next step will be the verification of the magnetic origin of this effect. This task is covered in the following chapter 7.

7. Commutation of the objective lens current of the TEM

Although the dichroic effect was measured clearly above the noise level (see chapter 6) and the experimental data agree well with the theoretical predictions (see chapter 2.4), the experimental evidence for the magnetic origin of this effect is still missing yet. As the sign of the dichroic effect depends on the orientation of the component of the magnetization in the direction of the electron path, the dichroic signal (as difference of two energy loss spectra taken at different positions in the diffraction plane) should change its sign when switching the magnetization direction of the specimen from parallel to antiparallel in respect to the electron beam.

The Tecnai microscope is equipped with a so called super-twin objective lens system. This means that the specimen is located in the center of the two-part pole piece of the objective lens. Each of the two parts of the pole piece is embedded in one objective lens coil and the magnetic flux leaks in the gap between the two parts and thus generates a magnetic lens with its optical center close below the specimen plane (compare left side of figure 7.5). By the use of this geometry, the specimen is always exposed to a magnetic field of up to 2 T (depending on acceleration voltage and defocus). Hence, considering micromagnetic simulations (see chapter 4), one can assume any specimen being extensively saturated in the magnetic field of the objective lens.

To assure the correlation of the specimen's magnetization and the dichroic effect, the current through the coils of the objective lens system of the TEM (and therewith at the same time the magetization direction of the specimen) can be reversed with a newly developed commutation unit (see chapter 7.2.1). According to theory, the dichroic effect should change sign by a reversal of the specimen's magnetization, as depicted in figure 7.1.

It is not self-evident that the TEM is still working satisfactorily after a commutation of the objective lens system, because the stray fields of all other (uncommutated) magnetic lenses may disturb the function of the objective lens. On contrary, also the stray field of the commutated objective lens system might degrade the imaging quality of the microscope as the focal length of a magnetic lens is finally defined by the magnitude of the magnetic field. To clarify this question, first of all the functionability of a magnetic electron lens is discussed in detail (chapter 7.1). Subsequently, the results of extensive studies on the imaging quality in reversed mode (chapter 7.2.2) and a hysteresis loop of the pole pieces of the objective lens is shown (chapter 7.2.3). Finally, the results of a dichroic measurement with reversed objective lens and a reference measurement are

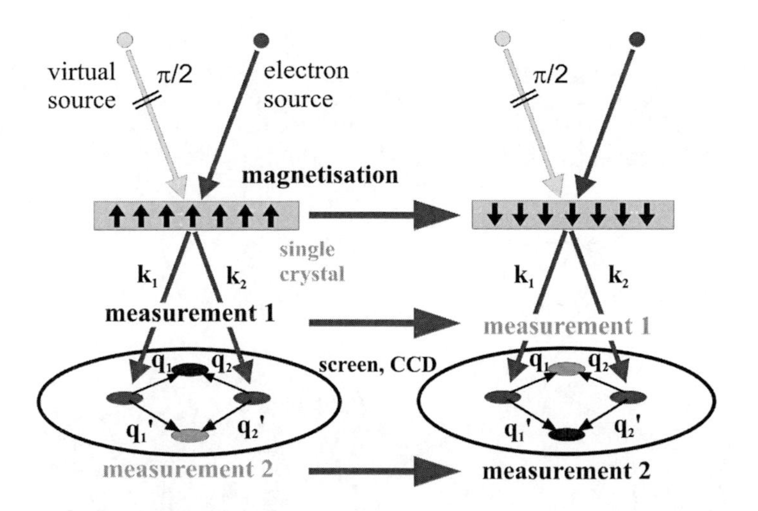

Fig. 7.1.: *Idea of the commutation measurement: The sign of the dichroic effect should change if the magnetization direction of the specimen is reversed (in this example from antiparallel to the electron beam to parallel to the electron beam). The experimental setup is equivalent to figure 3.5. Switched grayscales for measurement 1 and 2 represent the changed sign of the dichroic difference.*

presented (see chapter 7.3).

7.1. Functionability of electron lenses

Electrons in magnetic fields A reversal of the objective lens current is a drastic intervention in the imaging system of a TEM and can be done satisfyingly only if the principle of magnetic lenses for electron are well understood.

In a magnetic field, an electron in motion is deflected by the Lorentz force

$$\vec{F_L} = e\left(\vec{v} \times \vec{B}\right). \tag{7.1}$$

In equation 7.1, $e = -1,602 \cdot 10^{-19}\,\mathrm{C}$ is the elementary charge of the electron. If the magnetic field is *homogeneous* and the vector of movement $\vec{v}$ of the electron is not parallel or perpendicular to the magnetic field $\vec{B}$, the electron is accelerated on a helix like path. In this case, the vector of movement can be divided in a perpendicular component $v_\perp$ which leads to a circular path and in a parallel component $v_\parallel$ which leads to a linear movement in the magnetic field in this direction.

A first impression of the magnetic field acting as a lens for electrons is received by regarding the simplified case of two electron beam trajectories subtending in point P at the beginning of a magnetic field in z-direction that is assumed to be strictly spatially limited (see figure 7.2 a). The angle of incidence relative to the magnetic field is defined as α.

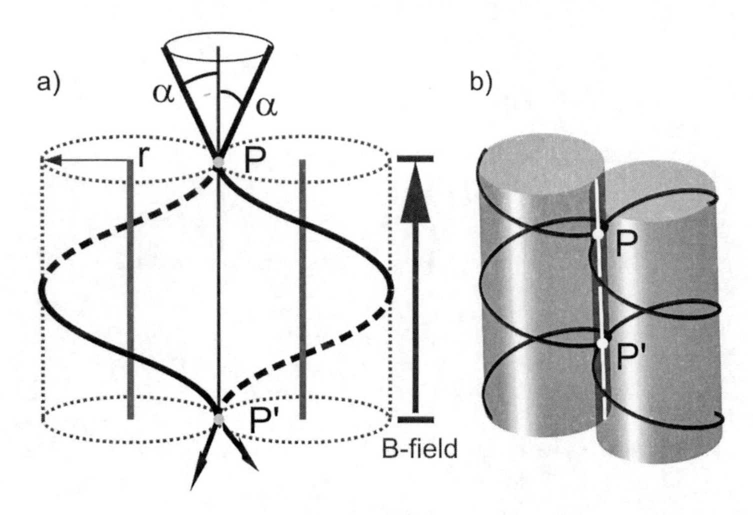

Fig. 7.2.: *a) In a homogeneous magnetic field B, electrons in motion are forced to helices due to the Lorentz force (equation 7.1). At a given value of B, the radius r_C depends on the component of the velocity perpendicular to the field (equation 7.2). The focal length $d_{PP'}$, defined as distance between the contact points P and P', depends on the component of the velocity parallel to the field (equation 7.4).*
b) Perspective drawing of two helices tangent to each other in the points P and P'.

For the following calculations, the electrons are relativistic[1]. In the present geometry, the component of the velocity perpendicular to the magnetic field is given by $v_\perp = |\vec{v}| \cdot \sin\alpha$ and the component parallel to $\vec{B}$ is given by $v_\parallel = |\vec{v}| \cdot \cos\alpha$.
According to [Rei97], the radius r_C of the circular movement caused by $v_\perp$ can be calculated using the equity of the Lorentz force $F_L = e \cdot v_\perp \cdot B$ and the centrifugal force $F_C = \frac{m \cdot v_\perp^2}{r_C}$:

[1] according to $v = c\sqrt{1 - \frac{1}{1+\frac{E}{E_0}}}$, at an acceleration voltage of 300 kV and a rest energy of $E_0 = 511\,\mathrm{keV}$ the relativistic velocity is $v = 0.85 \cdot c$

$$r_C = \frac{m \cdot v_\perp}{e \cdot B} = \frac{2m_0 E\sqrt{1+\frac{E}{2E_0}}}{eB} \cdot \sin\alpha, \tag{7.2}$$

with the electron energy $E = E_0 + E_{\text{kin}}$ ($E_{\text{kin}} = |U_{\text{acc}} \cdot e|$ is the kinetic energy of the electron after passing the acceleration voltage U_{acc}), the (relativistic) mass $m = \frac{m_0}{\sqrt{1-\frac{v^2}{c^2}}}$, the rest mass $m_0 = 9.1091 \cdot 10^{-31}$ kg, and the rest energy of an electron $E_0 = m_0 c^2 = 511$ keV.

The time t_C which the electron needs for one full circle is then

$$t_C = \frac{2\pi r_C}{v_\perp} = \frac{2\pi m}{eB}, \tag{7.3}$$

and only in the non-relativistic case ($m = m_0$) t_C is independent of the velocity of the electron. In this time t_C, the electron covers the distance $d_{PP'}$ in z-direction which can be calculated (using the cosine development $\cos x = 1 - \frac{1}{2}x^2 + ...$) to

$$d_{PP'} = v_\parallel \cdot t_C = |\vec{v}| \cdot t_C \cdot \cos\alpha = \frac{2\pi m\,|\vec{v}|}{eB}(1 - \frac{1}{2}\alpha^2 + ...). \tag{7.4}$$

Defining the prefactor $d_0 = \frac{2\pi m|\vec{v}|}{eB}$, this d_0 can be seen as focal length for small angles α (paraxial rays) and $d_{PP'}$ can be written as

$$d_{PP'} = d_0 - \Delta z. \tag{7.5}$$

The difference Δz caused by higher order terms of the cosine development can be seen as spherical aberration in analogy to the lens aberrations in light optics.
In this simplified case, it is obvious that a reversal of the magnetization direction has no influence on the focal length $d_{PP'}$ of this system as neither in equation 7.2 (for the radius of the helices) nor in equation 7.4 (for the distance between the tangent points of the helices) the sign of the z-component of the B-field appears, but only its absolute value.

Magnetic electron lenses For practical applications, the lens function must not require a certain incident angle of the electron beam as seen in the previous, simplified example. Thus, the precedent example has to be expanded to a non-homogeneous magnetic field.

The next realistic model for a magnetic lens is the rotational symmetric field of a short coil. Defining the z-axis like before as the optical axis, the magnetic field is composed at each position of a radial component B_r and an axial component B_z. An incident electron with components of the velocity v_r (perpendicular to the optical axis) and v_z (parallel to the optical axis) is deflected to ϕ-direction by the Lorentz-force (see figure 7.3 for the mechanism)

$$F_\phi = ev_r B_z - ev_z B_r. \tag{7.6}$$

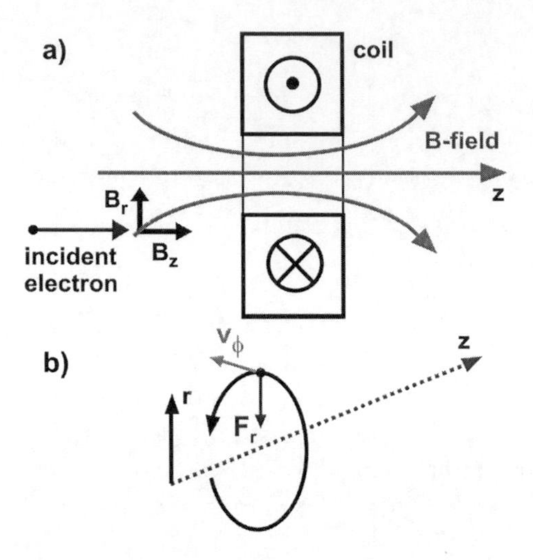

Fig. 7.3.: *a) Streamlines of the magnetic field B through a short coil. Due to the radial component B_r of the magnetic field, a velocity component v_ϕ is added to the electron's movement - even if it enters parallel to the optical axis (z).*
b) Due to the velocity component v_ϕ and the axial component B_z of the magnetic field, the Lorentz force F_r deflects the electron towards the optical axis.

Even if the electron enters parallel to the optical axis, it gets deflected on a screw like path in this inhomogeneous magnetic field. The component of the electron velocity in ϕ-direction v_ϕ now leads to a deflection towards the optical axis[2] as $v_\phi \perp B_z$ is always valid. The corresponding Lorentz-force can be written as

$$F_r = ev_\phi B_z. \tag{7.7}$$

Using the simplification that the extension of the lens field in z-direction is short and the distance between the electron and the optical axis thus rarely changes while the electron flies through the lens field[3], the focal length f of a short magnetic lens can be calculated to be [Nie93]

$$\frac{1}{f} = \frac{e}{8mU_{\text{acc}}} \int B_z^2 \text{d}z. \tag{7.8}$$

Similar to the previous model of the homogeneous magnetic field, also in this case the focal length only depends on B_z^2 and is thus independent of the orientation of the

[2]Magnetic electron lenses can only be converging but never dispersing lenses, because the deflection always appears towards the optical axis [Nie93].
[3]In the notation of equation 7.6, this means $v_r B_z \approx 0$.

magnetic field, respectively the sign of the current through the lens coil. To achieve short focal lengths of electron lenses (and thus large magnifications), the expansion of the magnetic field in z-direction must be short and the B_z-component of the field must be strong. This can be achieved by a concentration of the magnetic field by iron pole pieces with small gaps leaking the magnetic flux (see figure 7.5 for an example). In this gap, the magnetic field has rotational symmetry and the distribution of the z-component (the component along the optical axis) can be approximated by the so called *Glaser's Glockenfeld*:

$$B_z(z) = \frac{B_0}{1 + (z/a)^2} \tag{7.9}$$

and consequential ([Rei97])

$$B_r(z) = -\frac{r}{2}\frac{\partial B_z}{\partial z} \tag{7.10}$$

with a maximum field of B_0 in the center of the lens, $2a$ the full width at half maximum of the bell-shaped field and $r = 0$ the optical axis. Using this approximation, the equations of motion (and thus the calculation of parameters as focal length or image rotation) can be solved analytically as shown in [Rei97].

Due to the fundamental analogy to the examples with the homogeneous field or the simplified example with the field of a short coil specified above, it is comprehensible that also the focal length of magnetic electron lenses being based on the Glaser's bell-shaped field is invariant against a reversal of the direction of the z-component of the magnetic field (neglecting hysteresis issues as discussed in chapter 7.2.3 or stray field interactions as mentioned in chapter 7.2.1). The deflection of an incident electron wave, combined with a rotation of the image is sketched in figure 7.4 for both directions of the lens current.

7.2. Commutation performance

7.2.1. Commutation unit for the Tecnai TEM

Commutating a magnetic lens system with a magnetic field of up to 2 T and a ferrous polepiece with a weight of several kilograms can be dangerous - not only for the items of equipment, but also for the operator[4]. If the lens current would be switched off abruptly, the voltage that would be generated by self-induction might cause severe damage. Therefore an automatic commutation unit for a safe reversal of the objective lens current has been developed.

The supertwin lens system of the Tecnai TEM consists of two separate objective lens coils, respectively one at the top and one at the bottom position of the pole pieces.

[4]The energy stored in the magnetic field can induce a high voltage in the lens coils if the lens current is deactivated.

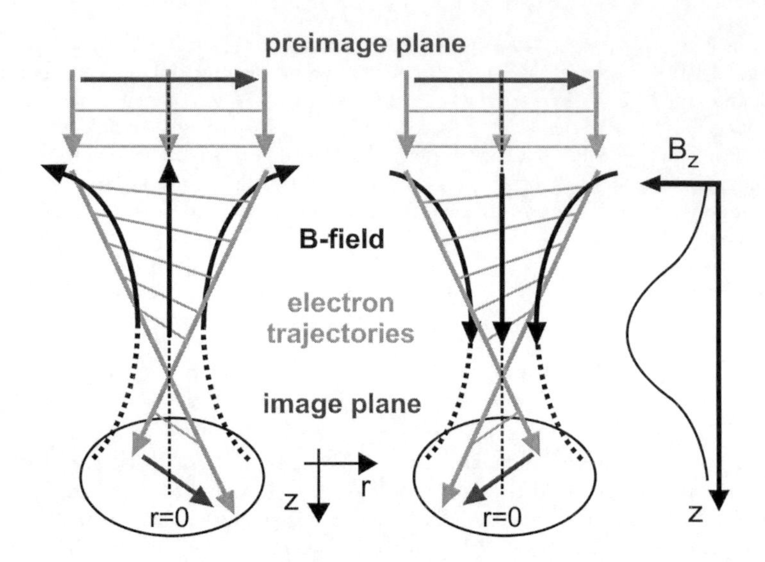

Fig. 7.4.: *Strongly simplified sketch of electron trajectories in a Glaser bell-shaped magnetic field. Whereas electrons along the optical axis ($r = 0$) pass unaffected, the deflection becomes larger for electrons with a growing distance r to the optical axis as the component of the B-field perpendicular to the electron's direction of movement increases. In reality, the off-axis trajectories are screw-like instead of straight lines. A change of the direction of the magnetic field (as visible from the left to the right side) changes the rotational direction of these helices but has no influence on the focal length of the lens.*

Both of them must be commutated for a reversal of the magnetic field at the specimen's location[5]. In the upper part of the pole piece, a small third lens coil with a gap above the specimen plane is included (see figure 7.5 for a cross section sketch). This so called minicondenser lens is activated or deactivated by a reversal of the current and thus a reversal of the magnetic flux generated by the coil. If the flux of the minicondenser lens coil is parallel to the flux generated by the objective lens coils, a magnetic field leaks the minicondenser lens gap and acts as magnetic lens (figure 7.5, left side). Otherwise, in the area of the minicondenser lens gap, the flux of the minicondenser lens coil annihilates with the flux of the objective lens coils and no magnetic lens appears (figure 7.5, right side). With an activated minicondenser lens, the TEM is operated in *microprobe* mode (constant illumination of a large area on the specimen). If the minicondenser lens is

[5]It is not possible to commutate the objective lens current right at the power supply! The current is monitored by the control system of the microscope - thus an inversion of the current causes a malfunction of the power supply. Only a commutation right at the objective lens coils is possible.

deactivated, the TEM is operated in *nanoprobe* mode (basically used for STEM[6]). The advantage of this switching technique is a constant thermal load on the pole piece, independent of the switching status of the minicondenser lens (and therefore less drift problems).

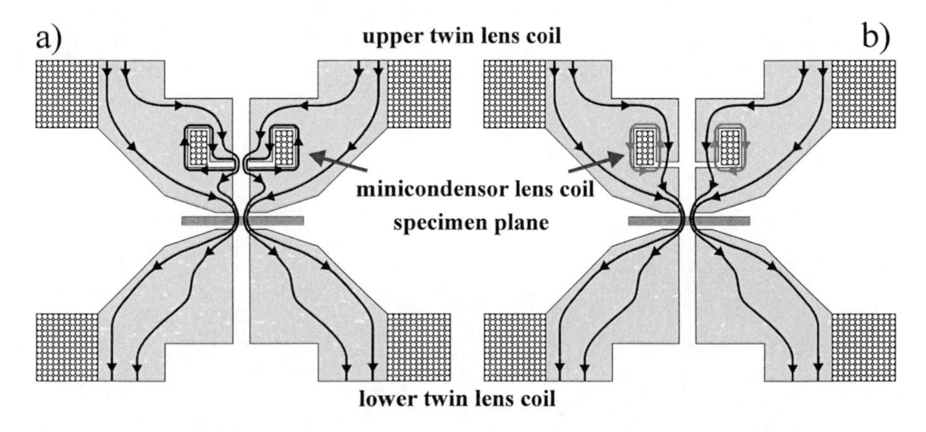

Fig. 7.5.: *The objective lens system of a Tecnai TEM consists of two objective lens coils at the ends of the pole pieces and a minicondenser lens coil in the upper part of the pole piece. The minicondenser lens is activated or deactivated - and the TEM thus operated in a) microprobe mode or b) nanoprobe mode - by changing the orientation of the magnetic flux of the minicondenser lens coil relatively to the flux of the objective lens coils. In each case, the specimen is penetrated by the the magnetic field of the objective lens.*

With reversed objective lens coils, one has to switch the microscope to *nanoprobe* mode to actually get *microprobe* mode (standard for TEM imaging). This works in principle, but the operation software allows only limited possibilities for the microscope alignment in *nanoprobe* mode. Due to this limitation, it is not possible to adjust identical imaging conditions in both states, normal current and reversed current of the objective lens. Thus, a reversal of the objective lens coils stringently requires also a reversal of the minicondenser lens coil[7]

Essential for the commutation is a safe shut down of the objective lens power supply. The microscope itself executes a shutdown of all lens currents if a problem with the cooling water is detected. This procedure can also be used for a shut down during a

[6]Scanning Transmission Electron Microscopy

[7]Technical problem of this solution is the missing plug contact of the minicondenser lens coil. A tap of the minicondenser lens coil can be found on the *"dual distribution board"* circuit board in the TEM cabinet.

commutation process[8].

Finally, after some feasibility studies using a Philips CM30 TEM[9], a fully automatic commutation unit with one-button operation has been developed (see block diagram 7.6 for an overwiew). The current state (normal/reversed) is displayed by an indicator light. After pushing the button, the lens current is deactivated (by simulating a lack of cooling water) and - after the electronics of the commutation unit measures the objective lens current to be zero - the lens coils (both objective lens coils and the minicondenser lens coil) are being commutated. Subsequently, the lens current is activated and the new state is displayed by the indicator light.

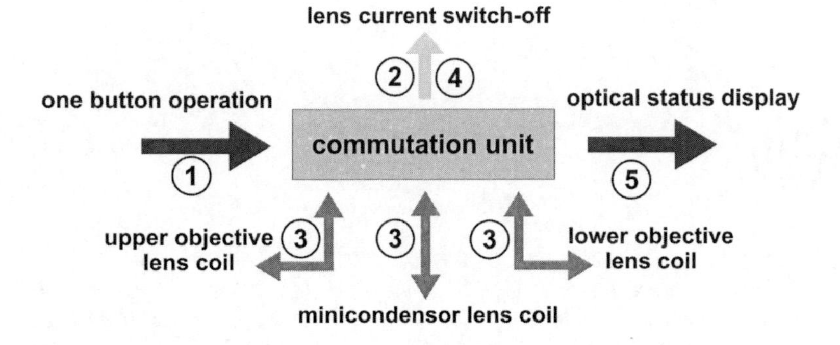

Fig. 7.6.: *Block diagram of the commutation unit:*
1) The unit is operated by one button, that starts a commutation process.
2) The lens current is switched off by simulating low cooling water.
3) The lens coils (two objective lens coils and the minicondenser lens coil) are being commutated.
4) The lens current is switched on again.
5) The new status is displayed by an indicator light.
Two prototypes of this unit are installed in Regensburg (Tecnai F30) and Vienna (Tecnai F20).

In reversed mode, some corrections for the alignment are necessary. These are in detail:

- focus - correction of the objective lens current
- intensity - correction of the intensity lens current

[8]The circuit to the floater reed contact has to be opened for a simulation of low cooling water. The circuit diagram for this function can be found on the *"objective lens control board"*

[9]The simpler construction of the CM30 allows a substitution of the original power supply by an external power supply for the objective lens.

- illuminated area - correction of the *usershift x-y* deflection lenses

The reason for these changes is the interaction of the stray fields of the commutated lens coils with the fields of other, uncommutated lens coils and contrariwise. The influence of a hysteresis of the objective lens system is discussed in chapter 7.9. In addition to this, the image of the specimen is rotated by a fixed angle, depending on the present objective lens excitation. This can be understood by the principle of a magnetic electron lens (the electrons move on helices in the magnetic field, compare chapter 7.1). If necessary, the rotation of the image can be manually corrected by adapting the current of the projection lens (P2) for the rotation of the image. As this correction of the rotation changes the magnification, the image size can be corrected by adapting the intermediate lens (IL) lens current[10].

7.2.2. Image quality

Taking the stray field interaction between the magnetic lenses into account, it is not as a matter of course that the imaging quality and the spatial resolution of the TEM is still expedient after a commutation of the objective lens system.
The spatial resolution is verified with a graphite specimen and the possibility for a reversal of the specimen's magnetization is checked with a Permalloy[11] specimen in Lorentz mode. The distance between the different layers of crystalline graphite particles is 0.34 nm. In figure 7.7 these fringes are clearly visible in reversed lens imaging mode. Thus, the spatial resolution with reversed lenses is at least 0.34 nm and in any case sufficient for EMCD measurements.

The second reference specimen is a circular ferromagnetic Permalloy dot[12] (with a diameter of 500 nm) - manufactured by electron beam lithography (EBL). In Lorentz mode, with an objective lens excitation of 10% and a tilt angle of 20 degrees around the y-axis, the resulting in-plane component of the objective lens field in x-direction is approx. $100mT$[13] and enough for a complete in plane saturation of the dot in x-direction as can be seen in the (false color) phase reconstructions of electron holographs in figure 7.8 (See [Voe99] or [Heu05] for details on electron holography). The phase of the incident electron wave is shifted on the one hand by the thickness dependent atomic potential of the specimen which is the reason why the dots are recognizable in their original shape. On the other hand, the phase is shifted by the magnetic induction in the specimen and the surrounding magnetic stray fields. After a reversal of the objective lens system, the orientation of the in-plane magnetization in x-direction has changed as it can be seen from the changed color gradient in y-direction.

---

[10] Both lenses are accessible in the *free lens control* or alternatively in the *TEMspy* menu

[11] $Fe_{81}Ni_{19}$

[12] Details on the magnetic properties of these structures can be found in [Lim06].

[13] according to [Ott01], for small excitations, the field of the objective lens is given by $\mu_0 H/\mathrm{mT} = 2.43 + 29.2 \cdot \frac{\text{OL excitation}}{\text{per cent}}$. A change of the remanent offset after the commutation which is not included in this formula is presumably.

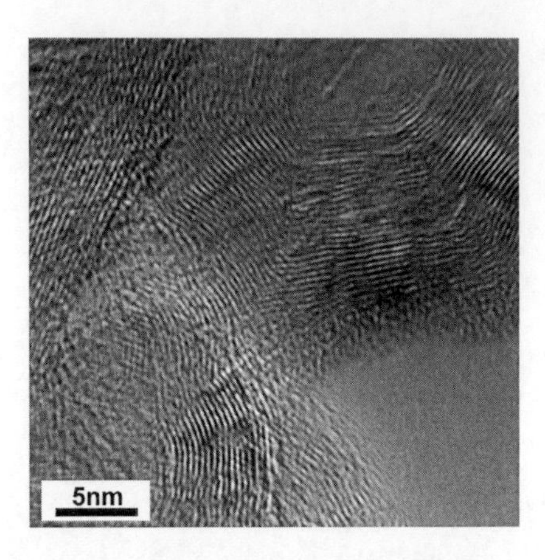

Fig. 7.7.: *Example for high resolution with reversed objective lens system: fringes in graphite with a distance of* 0.34 *nm can be resolved (SA imaging mode).*

On the one hand, this is a proof that it is smoothly possible to change the magnetization direction of a specimen which is needed for the desired EMCD experiment. On the other hand, this technique will become interesting also in other aspects of Lorentz microscopy, whenever the magnetization direction of a specimen has to be reversed. For example the recording of hysteresis loops or the thickness determination in electron holography require a reversal of the specimen's magnetization. Up to now, this was done by varying the tilt angle of the specimen which can have the disadvantage of a different focus plane at a certain area of the specimen, a changed crystal orientation or an altered effective thickness in the direction of the electron beam during the tilt process. By using the objective lens reversal instead, one has to be aware that these disadvantages are interchanged with a rotation of the image, which might in some cases be the lesser of two evils as it can be corrected by adapting the projection lens current in the free lens control as described before.

7.2.3. Hysteresis of objective lens pole pieces

Very important for the practical applicability of the lens reversal system is accurate information on the hysteresis of the objective lens pole pieces during a commutation cycle. The magnetic field generated by a magnetic lens is a proportion for the electron-optic refraction power of the lens. Thus, a change of the magnetic field is equivalent to a change of the focus plane of the lens system.

To assess these influences, the magnetic field can be measured in the specimen plane with a particular specimen holder, equipped with a Hall sensor instead of an electron transparent specimen ([Ott01], compare also section 4.1.

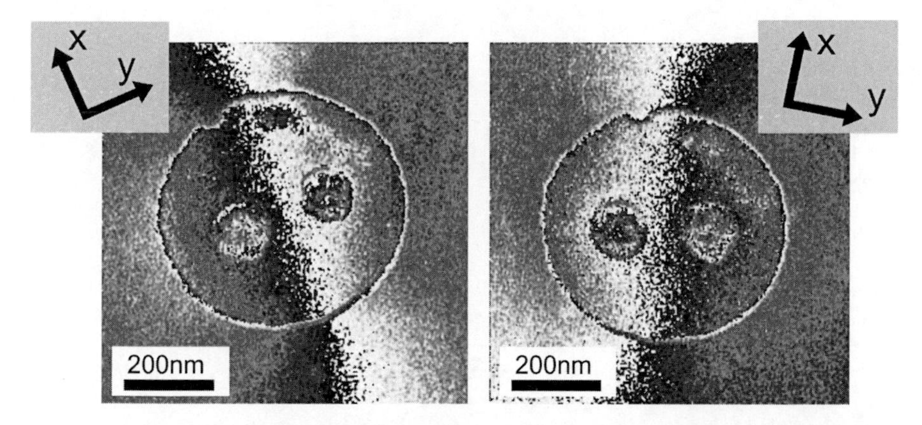

Fig. 7.8.: *False color phase reconstructions of electron holographs of a ferromagnetic Permalloy dot with normal objective lens current (left side) and reversed objective lens current (right side). In Lorentz mode, with an in-plane field of approx.* $100\,mT$*, generated by a* 10% *excited objective lens and a tilt of 20 degrees around the y-axis, a change of the grayscales gradient in y-direction and thus a change of the orientation of the in-plane magnetization in x-direction is visible from the left to the right hologram.*
The two holes and the damage in the annulus have no importance for this experiment, but they relieve the correlation of the two images as they are rotated against each other at 27 degrees due to the reversal of the objective lens.

The following lines briefly show the course of action in this section to prevent confusion.

- First of all, at an acceleration voltage of 300 kV, a complete hysteresis loop is recorded.
- The difference between the two branches of the hysteresis loop is calculated and an adequate function is fitted to allow further interpolation.
- From this data, the absolute value $\Delta U(OL)$ and the gradient $G(OL)$ of the difference signal can be calculated for a - in workaday life typical - lens excitation of $OL = 90\%$.
- The quotient $\Delta OL_{\text{eff}}(OL) = \frac{\Delta U(OL)}{G(OL)}$ gives a value for the hysteretic change in per cent of the lens excitation.
- Again for a lens excitation of 90%, the change on the relative defocus (Δf_{rel} in μm) per change of the lens excitation (in per cent) is read off the TEM user interface.
- The maximum possible defocus after one commutation cycle is determined by $\Delta f_{\text{eff}} = \Delta OL_{\text{eff}} \cdot \Delta f_{\text{rel}}$.

- Finally, this result is compared with experimental observations of the defocus after one commutation cycle.

One complete hysteresis loop is shown in figure 7.9 from 100% objective lens excitation to 0% excitation, then, after commutation, from 0% excitation to -100% excitation and back to 100% excitation (again with a reversal at 0% excitation). Within the error bars, no hysteresis (remanence or coercive field) is detectable. Assuming the relative error between the two branches of the hysteresis loop significantly smaller than the absolute error (this is lifelike as for example at stable surrounding conditions the error of measurement of the multimeter will not change sign within some minutes), it is interesting to regard the difference signal between the two branches. This difference (in the graph it is amplified by a factor of 100) shows the typical hysteresis of a magnetically soft material.

The fit of a quadratic function allows an interpolation of the hysteresis between the two branches (+100% to -100% and back to +100%) :

$$\Delta U = p_1 + p_2 \cdot OL + p3 \cdot OL^2 \qquad (7.11)$$

with OL - objective lens excitation in per cent, ΔU - difference between the two branches in Volt[14] and the parameters[15]

$$\begin{aligned} p_1 &= (2,05 \pm 0,05) \cdot 10^{-2}\,\mathrm{V} \\ p_2 &= (-4,10 \pm 7,06) \cdot 10^{-6}\,\mathrm{V} \\ p_3 &= (-1,93 \pm 0,12) \cdot 10^{-6}\,\mathrm{V} \end{aligned}$$

For the analysis, additionally the 95% prediction bands (95% of all measured points are within these two bands) are plotted. The choice of a quadratic function has no physical background and is admittedly no adequate approximation, especially not around the point -100% and $+100\%$. Here, the curve should converge to zero instead of crossing the axis of abscissae. But as 95% of the data points remain between the prediction bands, the approximation is good enough for a rough estimation of the defocus caused by hysteresis effects. It is also probable that the polepieces are not completely saturated at a lens excitation of 100%, because the difference is still noticeable above zero at 100% excitation. As a stronger excitation of the objective lens is not possible due to technical limitations, one has to deal with this interval from -100% to $+100\%$.

The gradient $G(OL)$ of the hysteresis loop at $OL = 90\%$ objective lens excitation is $G(90\%) = 0.016 \frac{V}{\%OL}$ (in this accuracy not depending on the chosen branch of the loop). At this excitation, the corresponding difference signal between the two branches of the hysteresis loop can be calculated with equation 7.11

[14]... arbitrary reference due to the signal amplification of the Hall sensor chip

[15]keep in mind: the plot of ΔU in 7.9 is magnified by a factor of 100)

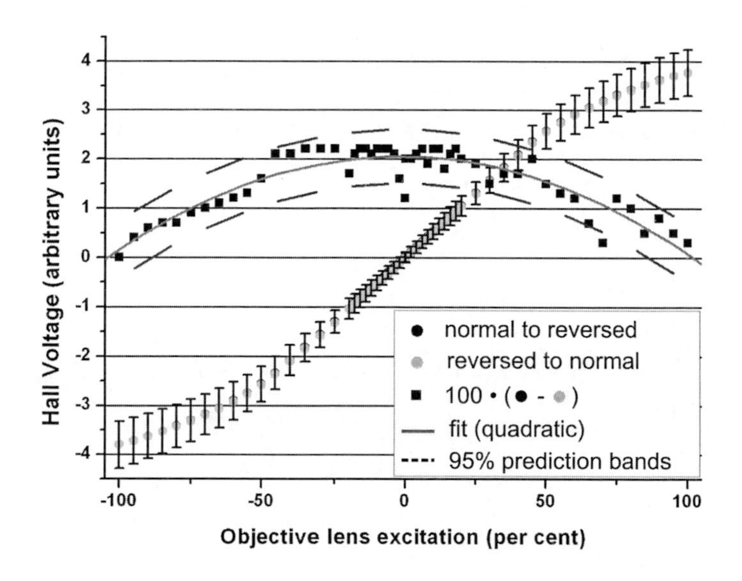

Fig. 7.9.: *Hysteresis loop of the magnetic field at the specimen's location in the gap between the objective lens pole pieces (Tecnai F30, 300 kV). As described in section 4.1, the error comes from the nonlinearity of the Hall sensor at high fields. Whereas within the absolute error bars (only added to the black points (from 100% to -100% excitation), which are almost completely covered by the grey ones (from -100% to +100% excitation)), no hysteresis is detectable, the difference of the two branches (black squares, here amplified by a factor of 100) shows the typical hysteresis difference [Tip94] of a magnetically soft material. For the analysis, a quadratic fit and the 95% prediction bands are plotted. The voltage is proportional to the magnetic field and is intentional not converted to keep the absolute error small. A voltage of 4 V is equivalent to a magnetic field of approx. 2.1 T [Ott01].*

$$\Delta U(\mathrm{OL} = 90\%) = (4.5 \pm 5.5) \cdot 10^{-3}\,\mathrm{V} \tag{7.12}$$

The error limit only results from the width of the 95% prediction bands. Due to physical considerations, for ΔU and the following results a value smaller than zero is not possible. So, for practical reasons, in each case the error interval is limited by zero. Equation 7.12 describes a shift of the effective magnetic field at the position of the specimen which (assuming a hysteresis free system) corresponds to a change in the objective lens excitation of

$$\Delta \mathrm{OL}_{\mathrm{eff}}(90\%) = \frac{\Delta U(90\%)}{G(90\%)} = (0.28 \pm 0.34)\%. \tag{7.13}$$

In the common operating mode of the TEM (SA (selected aperture) mode or SA diffraction mode) which is also used for dichroic measurements, the excitation of the objective lens is (depending on the current defocus) about 90%[16]. According to the defocus display of the Tecnai user interface, around an excitation of 90%, a change of the objective lens excitation of $\Delta OL = 1\%$ effects a shift of the focus plane of $\Delta f = 23.2\,\mu$m. Thus,

$$\Delta f_{\mathrm{rel}} = 23.2\,\frac{\mu\mathrm{m}}{\%\mathrm{OL}}. \tag{7.14}$$

This is equivalent to an effective shift of the focus plane due to hysteresis of

$$\Delta f_{\mathrm{eff}} = \Delta OL_{\mathrm{eff}} \cdot \Delta f_{\mathrm{rel}} = (0.28 \pm 0.34)\% \cdot 23.2\,\frac{\mu\mathrm{m}}{\%} = (6.5 \pm 7.9)\,\mu\mathrm{m} \tag{7.15}$$

after a complete commutation cycle at an objective lens excitation of 90% (assuming the pole pieces once saturated beforehand and neglecting a change in the hysteresis loop due to not completely exciting the objective lens to −100%). In the worst case, this defocus of $(6.5 \pm 7.9)\,\mu$m remains after a commutation cylce as remanent offset of the focal plane. Compared to the typical defocus in Lorentz microscopy[17] of $10\,\mu$m to $20\,\mu$m, this offset appears non-critical and can easily be corrected by the use of the focus adjusting knob.

The specimen holder itself is manufactured of brass and may contain an undefined percentage of other - possibly ferromagnetic - elements. Thus it is possible that also the specimen holder contributes to the detected hysteresis to an amount that is not to quantify. The result has to be evaluated against this background that in the worst case, all of the hysteresis is generated by the pole pieces of the objective lens, but probably, the hysteresis generated by the pole pieces is lower due to a possible hysteresis of the specimen holder. The defocus after a commutation of the objective lens current was also estimated experimentally. In practice, a correction of less than $5\,\mu$m was necessary after the recommutation.

In any case, as the difference between the two branches approaches zero at an objective lens excitation of 100% respectively −100%, the effect of a hysteresis of the pole pieces can be reduced by a brief excitation of the objective lens system to 100% after a commutation measurement.

[16]keeping all other parameters of the TEM fixed, a variation of the objective lens current causes a shift of the focus plane in addition with a marginal rotation of the image and a small change of the magnification

[17]Lorentz microscopy was not used for EMCD experiments yet - it is only used here for a comparison as Lorentz microscopy usually deals with defocus values.

7.3. EMCD commutation measurements

Before comparing two dichroic spectra, taken at normal and reversed objective lens currents, it must be ensured that the commutation process itself has no influence on the energy loss spectrum of a specimen. For this reason, the spectrum of the $L_{3,2}$ edge of a nickel specimen (the identical one from the measurement shown in chapter 6) was taken at the same location on the specimen once in normal mode and once with a reversed objective lens system. Within the noise, no difference was detectable (see figure 7.10).

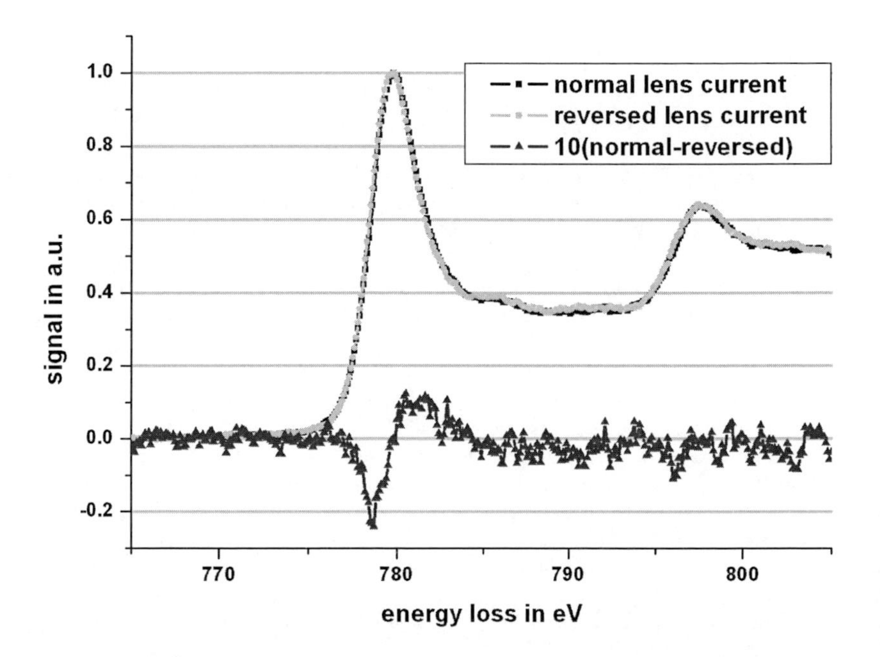

Fig. 7.10.: *Two spectra of a nickel specimen, taken at the same position on the specimen, but with different signs of the objective lens current. The difference signal is heightened by a factor of 10 and shows a strain between the two spectra (due to a drift of the energy dispersion). The area under the negative peak (below 780 eV) and the positive peak (above 780 eV) is similar. Thus, no dichroic signal is visible within the noise - a misleading effect due to the lens commutation can be excluded.*

The comparison of two dichroic spectra pairs, one taken with normal objective lens current and the other taken in reversed mode, was done by the ChiralTEM project partner in Vienna[18]. Although nickel was determined as best applicable specimen for EMCD, the comparative measurement was performed on a cobalt specimen. A cobalt specimen reaches the maximum dichroic signal at a larger specimen thickness and is thus less vulnerable to the electron beam (for the comparative measurement, two times two spectra have to been taken at the identical position on the specimen - without reducing the specimen thickness between the four measurements which is a known problem on nickel specimens, compare chapter 6).

When commutating the objective lens current, the diffraction pattern is rotated relative to the uncommutated case. Using the *diffraction shift method* (compare section 3.3.1), the rotation does no need to be compensated as the two positions for the measurements are selected by a shift of the diffraction pattern relative to the spectrometer entrance aperture. One only has to be sure not to confuse the zero spot and the G-spot after the rotation[19]. If the *spectrum spread method* (compare section 3.3.2) is used instead, the 0-G-axis is predefined to be perpendicular to the energy-loss axis on the CCD. In that case, the rotation of the diffraction pattern has to be corrected by adjusting the projection lens (P2) and adapting the magnification with the intermediate lens (IL) afterwards in order to have the same imaging conditions after the commutation. For the measurements in Vienna, the spectrum spread method was chosen, because the electron irradiation on the specimen is smaller compared to the diffraction shift method.

The result of the measurement is shown in figure 7.11. In any case, the sign of the dichroic signal changes after a commutation of the objective lens and thus the magnetic origin of the dichroic signal is confirmed. No metrological reason can be given for the different size of the dichroic signal in both measurements. The dichroic signal in cobalt is small[20] anyway and the detected dichroic signal is anyhow volatile in practice. Already a small drift of the specimen between the two measurements or some bending due to the heating caused by the electron beam can change either the specimen thickness or the tilt angle and thus affect the dichroic signal. It is also known that the alignment of the energy filter is very sensitive on any change of external magnetic fields. Of course, also the reversal of the stray fields of the objective lens after a commutation will degrade the spectrometer alignment.

[18]Due to technical problems, the Tecnai TEM was predominantly not available in Regensburg in 2007/2008.

[19]The zero-spot and the G-spot can be distinguished by expanding the beam in diffraction mode. Only in the zero-spot, the hole in the specimen appears bright - in the G-spot and all other diffracted spots, the hole appears dark.

[20]maximum 16.3% at the L_3-edge according to theoretic predictions, see section 2.4.

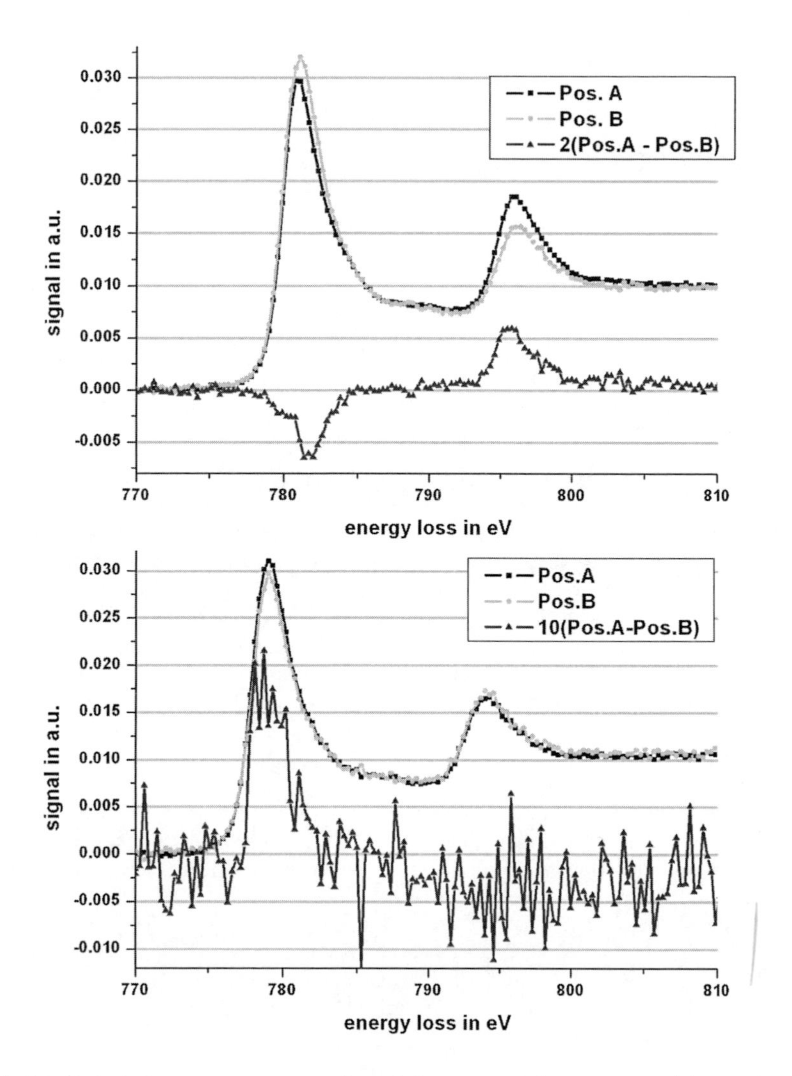

Fig. 7.11.: *Two dichroic spectra pair of a cobalt specimen ($t = 20 \pm 2\,nm$)) at the cobalt $L_{3,2}$ edge (spectra taken by Michael Stoeger-Pollach, Vienna). The top spectrum was taken with the normal objective lens current and the bottom spectrum was taken with reversed lenses. A significant change of the sign of the dichroic effect is visible from top to bottom, although the difference between the two spectra taken in reversed mode is smaller as the alignment of the imaging filter is jammed due to the reversed stray field of the objective lens.*

8. Outlook

In this chapter, some prospects for future EMCD experiments are demonstrated. At the beginning of the ChiralTEM project, the probability for a successful experimental verification of the dichroic effect was not foreseeable. The same statement is now true for every of the experiments suggested in this chapter. While experiments on rare earth elements (section 8.1) are rather promising, the chances of success for a different experimental setup that works without a single crystal specimen (section 8.2) and for EMCD in Lorentz mode (section 8.3) are unsure.

8.1. EMCD on the rare earth element Gd

After the verification of EMCD on iron, nickel, and cobalt specimens [Rub07a], two possible directions for further developments are possible - tuning into a better signal to noise ratio with a concurrent increase of the spatial resolution or approaching different magnetic materials. As the Vienna group actually follows the first of the two directions [Scha08], [Hou07] the decision in Regensburg was to explore new materials. A deficient availability of the TEM in Regensburg in the last phase of the ChiralTEM project prevented extensive investigations, for example of garnet specimens as described in chapter 5. Perhaps, the experimental verification of EMCD on a gadolinium specimen would be even more interesting.

Any ferromagnetic specimen is only ferromagnetic at temperatures below the Curie temperature, T_C. Beyond this point, if $T > T_C$, the specimen becomes paramagnetic due to thermal excitation, which disturbs the order of the spins. Typical values for T_C according to [Kit02] can be found in table 8.1.

Element	T_C
Fe	1043 K
Ni	627 K
Co	1388 K
Gd	292 K
Dy	88 K

Table 8.1.: *In contrast to iron, nickel and cobalt with Curie temperatures far beyond room temperature, the Curie temperature of the rare-earth elements gadolinium and dysprosium is considerably lower.*

The Curie temperature for iron, nickel and cobalt is far beyond room temperature. In contrast to these elements from the 4th main group of the periodic table, the ferromagnetic rare earth elements (6th main group), such as gadolinium or dysprosium, have a considerably lower T_C.

For example a Curie temperature of 292 K (Gd) can be adjusted by using a liquid nitrogen specimen holder. With the available double tilt holder *Gatan Model 636 - Double Tilt LN_2 (Liquid Nitrogen) Cooled Specimen Holder*, a specimen temperature in a range between 130 K and 393 K is adjustable[1] [Gat92].

Thus, the ferromagnetism in Gd can be switched "on" and "off" by changing the temperature of the specimen. Following the commutation of the specimen's magnetization (chapter 7), this would be a second, conclusive verification of the magnetic origin of the dichroic effect.

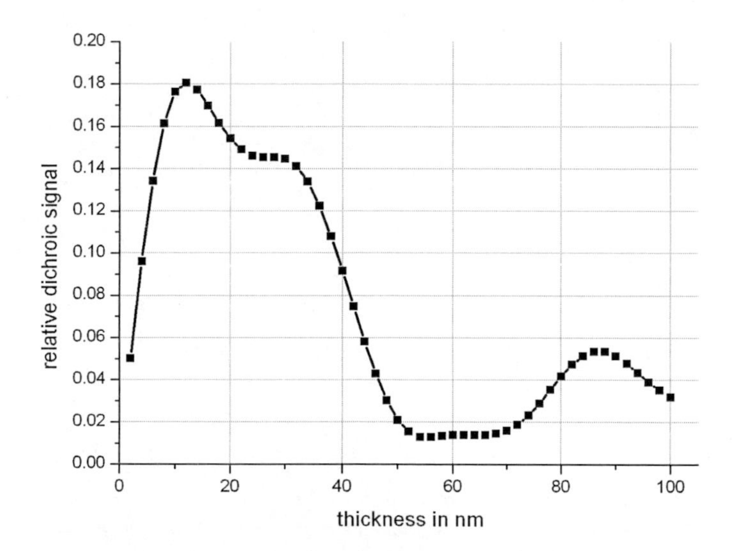

Fig. 8.1.: *Predicted relative dichroic signal at the gadolinium M5-edge (assuming Gd in* (110) *orientation and an acceleration voltage of* 200 *kV). The signal reaches a maximum of* 18% *at a film thickness of* 12 *nm and is larger than* 10% *in a range of* 6 *nm to* 38 *nm. Data provided by Jan Rusz, Uppsala.*

[1] With a double tilt liquid helium specimen holder, even Dysprosium might be investigated. But the available cooling holder has no possibility for double tilt and is thus not suitable for EMCD experiments with the current experimental setup.

Different to Fe, Ni and Co, the strongest dichroism in Gd is expected at the $M_{4,5}$ edges [Rus08b]. Jan Rusz calculated the relative dichroic effect versus the specimen thickness as described in section 2.4. The result is shown in figure 8.1. A maximum relative dichroic signal of 18% at the M_5-edge in a gadolinium film of 12 nm is predicted. The thickness range with a relative dichroic signal larger than 10% reaches from 6 nm to 38 nm and is thus larger than the corresponding range for Fe, Ni or Co (remember table 2.2).

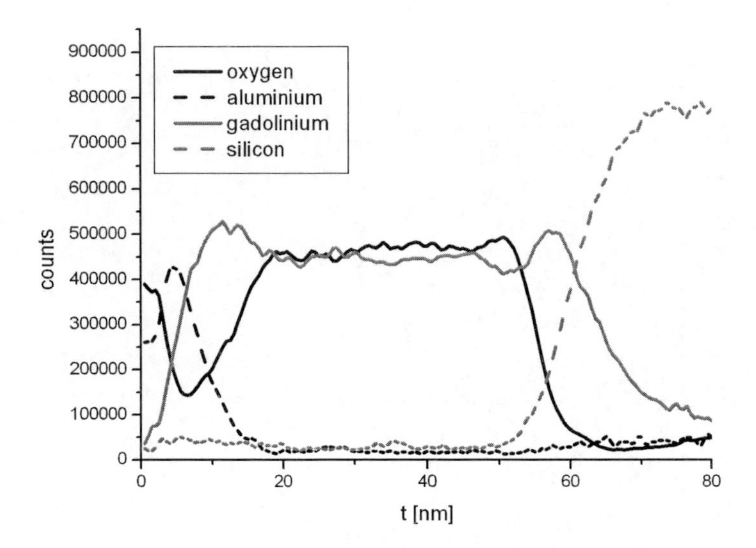

Fig. 8.2.: *Result of an Auger electron spectroscopy (AES) of a gadolinium layer grown on silicon substrate in a MBE. Clearly visible is (from left to right) the aluminium capping layer for oxidation protection, the gadolinium layer (approx. 50 nm) and below the silicon substrate. The constantly high oxygen concentration along the gadolinium layer gives a hint for an in-situ oxidation during the vacuum deposition of the gadolinium [Her06].*

Within the scope of a diploma thesis [Her06], it was tried to grow a Gd layer on a Si_3N_4 membrane [Twe08], using MBE[2]. Although evaporating in high vacuum and using a 5 nm aluminium capping layer for oxidation protection, AES[3] measurements showed

[2] **M**olecular **B**eam **E**pitaxy

[3] **A**uger **E**lectron **S**pectroscopy, compare [Pla92] for details on this technique.

a complete oxidation directly after the vacuum deposition of the layer system (compare figure 8.2). Consequently no magnetic domains were observed in these specimens.

This oxidation problem can be solved by a bulk preparation of gadolinium using the techniques polishing, dimpling, and ion-etching as described in section 5.3.1. As a first preparation step, Gd disks with a thickness of 0.75 mm were cut out of a polycrystalline bulk by wire-cut EDM[4]. A Fresnel Lorentz micrograph shows magnetic domains in the area directly at the hole (compare figure 8.3) - a confirmation of the ferromagnetic property of the specimen. It is not possible to evaluate how far an oxidation - starting from the surfaces - intrudes the specimen. As a capping layer is missing, the oxidation process will proceed until a complete oxidation is reached[5].

Fig. 8.3.: *Lorentz micrograph of a gadolinium specimen. For example in the marked area, the walls between magnetic domains can be seen as alternating bright and dark lines. The presence of these domains is a confirmation of the ferromagnetic property of the investigated gadolinium specimen.*

For the dichroic measurement, the lowest possible temperature was adjusted for two reasons: On the one hand, the saturation magnetization at temperatures only slightly below the Curie temperature is much smaller than far below T_C. On the other hand, a temperature above the lowest possible temperature is controlled antagonistically by the use of a heating filament. This technique leads to periodic fluctuations in the temperature

[4]**E**lectrical **D**ischarge **M**achining

[5]After a storage time of some weeks in the vacuum of the exsiccator, no magnetic domains have been visible in Lorentz mode.

(with a period of a few seconds) and therefore to a perturbing periodic drift of the specimen.

The crystallites of the available polycrystalline bulk material were too small for a systematic orientation of the diffraction pattern. Thus, it was not possible to make reproduceable EMCD measurements on gadolinium. A pair of spectra, that gives a hint for an existing dichroic difference is shown in figure 8.4. Currently, a new source for the vacuum deposition of gadolinium is installed in Regensburg within the scope of a diploma thesis [Koe08]. Having specimens with large crystallites, a well defined specimen thickness and a capping layer against oxidation, future EMCD experiments on gadolinium are promising.

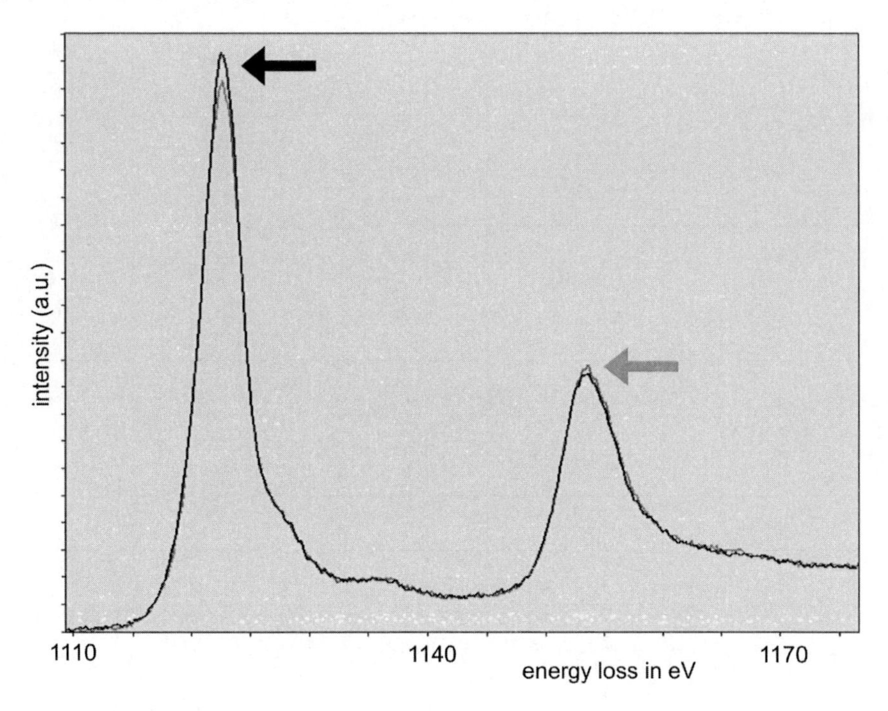

Fig. 8.4.: *These two energy loss spectra, taken at the $M_{5,4}$-edge of a bulk prepared gadolinium specimen in a liquid nitrogen specimen holder, seem to show a small dichroic difference. In fact, the problems concerning oxidation, crystallite size and spectrum drift were uncontrollable and the results hardly reproducible. Thus, these spectra may be taken only as a hint for EMCD in gadolinium.*

8.2. New experimental setup for EMCD

The practicability of EMCD would be increased drastically if the requirements on the specimen were reduced. Up to now, the specimen has to act as a beam splitter and thus has to show large crystallites. As a consequence of the diffraction theory, the intensity of the dichroic signal strongly depends on the specimen thickness (compare section 2.4). This disadvantage might be resolved by a different experimental setup that uses no longer the specimen itself as a beam splitter.

Within the ChiralTEM project, a setup using an electrostatic biprism in the condenser aperture plane was tested, but did not lead to satisfying results yet [For05]. Currently, the combination of a double-hole aperture with an electrostatic phase-shifter in the condenser aperture plane is discussed within a diploma thesis at the University of Regensburg [Has08]. This setup would use a Boersch phase plate [Mat96] in one of the holes to adjust a relative phase shift between the two partial waves through the two holes. According to Fourier optics [Spe03], for a given radius and distance of the holes, a value for the phase shift can be found to receive a maximum in the interference pattern that has a phase shift of $\frac{\pi}{2}$ between the two coherent partial waves. An exemplary combination can be seen in figure 8.5. In a very rough estimation, the contribution to the dichroic signal along the x-axis is set to be +100% for a phase shift of $\frac{\pi}{2}$ and −100% for a phase shift of $-\frac{\pi}{2}$ with a linear gradient in between. Doing so, the usable ratio of the signal can be estimated to be more than 70% which would be rather promising. First experiments with a double-hole condenser aperture in STEM-mode are pending.

8.3. EMCD in Lorentz mode

If EMCD wants to draw level with XMCD, the visualization of magnetic domains must be facilitated also by EMCD. Therefore, the Lorentz mode might be useful. When the TEM is operated in Lorentz mode, the objecitve lens is not needed for the imaging of the specimen plane. Thus, it can be deactivated completely to have the specimen plane almost free of any magnetic fields (compare section 3.1).

To have at least a part of the specimen's magnetization in the direction of the electron beam, specimens with a strong perpendicular anisotropy are required. Possible specimens are for example cobalt [Aum05], yttrium-iron-garnets (YIG) [Bec02] or different multilayer systems such as Fe/Au or Fe/Tb [Koe02]. EMCD in Lorentz mode has not been tried yet. One has to expect the following problems when making EMCD measurements in Lorentz mode:

- **Worse imaging conditions in Lorentz mode** - As the electron beam is not converging below the (not excited) objective lens in Lorentz mode, the intensity is reduced by a spatial limitation due to the limited diameter of the objective lens itself (compare figure 3.1).

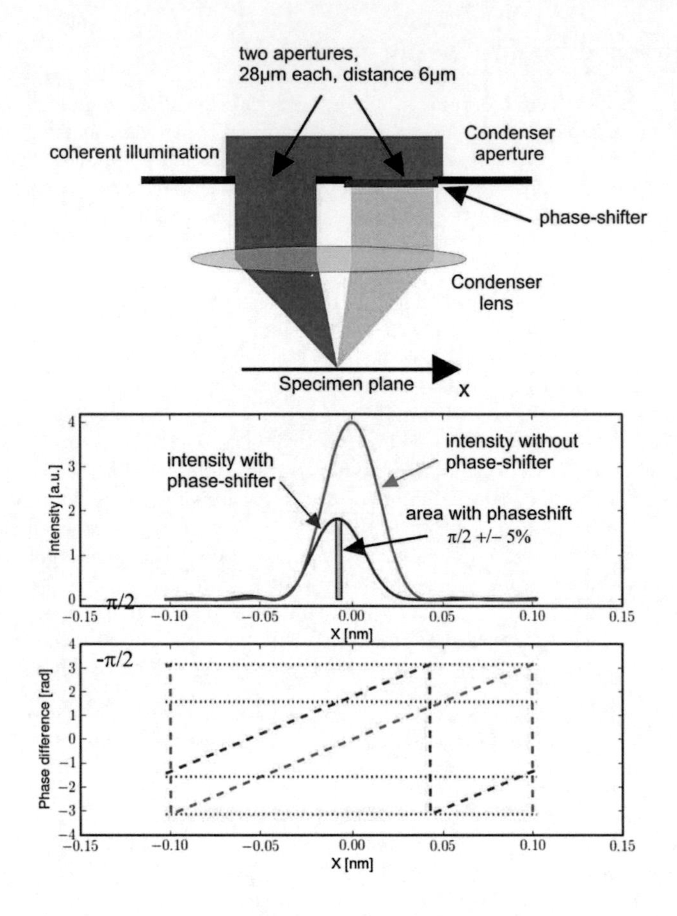

Fig. 8.5.: *The incident, coherent electron beam is divided by a two-hole (radius 28 µm) condenser aperture. One partial wave is phase-shifted relative to the other. For an exemplary focal length of the condenser lens (STEM-mode) of 1 mm and an electron wavelength of 4 pm, the ideal condition (phase shift of $\frac{\pi}{2}$ at the maximum of the intensity in the specimen plane) can be found for a phase shift of 0.57π between the two partial waves (in the condenser aperture plane). In this case, a phase shift of $\frac{\pi}{2}$ predominates the peak of the intensity [Has08].*

- **Different diffraction conditions** - Due to different lens excitations, the camera length is considerably longer in Lorentz mode. As the (largest) spectrometer entrance aperture has a fixed value, the contributing relative area of the diffraction pattern is reduced. This will become a problem concerning the signal to noise ratio.

- **Magnetic domains not visible during EMCD measurement** - Using the current experimental setup, the TEM has to be switched to the diffraction mode for an EMCD measurement. Thus it is difficult to make an EMCD measurement at a certain position on the domain pattern.

- **Reduced magnetization in the direction of the electron beam** - According to the micromagnetic simulations in chapter 4, the surface magnetization is always parallel to the specimen plane. Thus, the *effective* layer thickness (with a magnetization perpendicular to the specimen plane) will be reduced.

Summarizing all these aspects, EMCD in Lorentz mode will be no easy task. Alternative experimental setups might help to solve some of the problems described in this section.

At the very beginning of the ChiralTEM project, it was discussed to manipulate the specimen's magnetization either by a reversal of the objective lens or by a dedicated specimen holder. As already known from chapter 7, the lens reversal was the preferred choice. Nevertheless, a pulsed field specimen holder was constructed in Regensburg. The tip of this specimen holder is shown in figure 8.6. The specimen is inserted in a slit between two conduction circuits. This circuit is connected by two magnet wires with a diameter of 2 mm each. If a current pulse is sent through the circuit, a magnetic pulse (to a large extent perpendicular to the specimen plane) will force a remanent change of the magnetic domain structure of the specimen. For future EMCD experiments in Lorentz mode, this development might be useful. One big problem is the missing double-tilt possibility that facilitates the selection of a two beam case in diffraction mode.

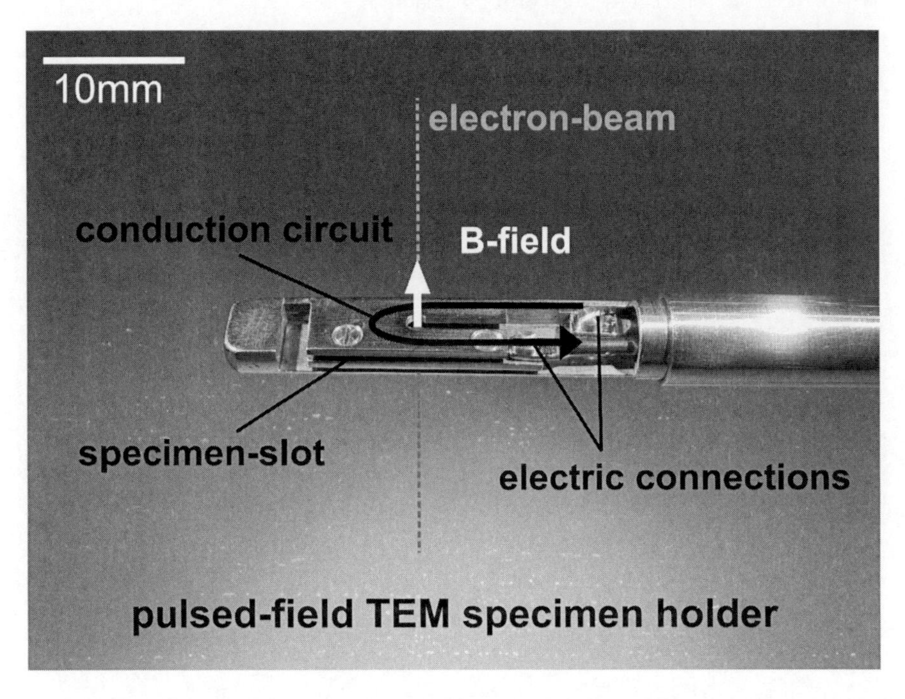

Fig. 8.6.: *Image of the tip of the pulsed-field specimen holder. The specimen is inserted between the two conduction circuit planes, electrically isolated by mica or Teflon plates and fixed with two screws. A current pulse through the conduction circuit planes generates a magnetic field pulse, which is at the position of the specimen is mainly parallel to the electron beam.*

9. Summary

One of the main tasks of the ChiralTEM project was an experimental verification of Electron (Energy Loss) Magnetic Chiral Dichroism (EMCD), the electron counterpart to X-ray magnetic circular dichroism (XMCD). In the short time from the end of 2004 until the middle of 2007, this task was not only achieved, but even excelled by far. In Vienna, EMCD was detected in iron, cobalt and nickel specimens clearly above the noise level. The theoretic predictions of the dichroic signal were reproduced to a large extent in the experiments. And with the confirmation of EMCD and XMCD on one single specimen, even a direct experimental comparison was done [Scha06].

Within this European collaboration, the main task of the University of Regensburg was the assortment, preparation and characterization of suitable specimens with clearly known magnetic parameters. Therefore, the magnetic properties, especially the external field required for a complete saturation (parallel to the external field) of iron, nickel and cobalt specimens, was determined using micromagnetic simulations. Lots of different preparation techniques, such as different vacuum deposition techniques, bulk preparation by ion etching and electrochemical etching were evaluated regarding their usability to produce ideal specimens for a verification of EMCD. Every element and every preparation technique has its advantages and disadvantages - some of them don't even show up until a certain combination of element and technique. Of course, these results can be discussed extensively, but an advice for an uncomplicated measurement of EMCD seems to be a nickel specimen, prepared by an electrochemical etching process.

A second task was to find a way for an in-situ manipulation of the specimen's magnetization. It was decided to follow the approach of a complete reversal of the objective lens current and thus forcing a complete reversal of the specimen's magnetization, which is (in most cases) completely saturated in the field of the objective lens. Automatic commutation units were developed and installed in Vienna and Regensburg. Finally, the magnetic origin of the dichroic effect was demonstrated unquestionably.

Evaluating the applicability of today's EMCD as a tool for magnetic investigations, it has to be noticed that the strict requirements on the investigated specimen rule out almost any practical application. Only single crystal specimens can be used as the specimen itself has to act on the one hand as a beam-splitter and on the other hand as magnetic layer. Thus, at a given element, the intensity of the dichroic signal depends not only on the specimen thickness, but also on the crystal orientation. Probably, future experimental setups will be able to circumvent these requirements for well defined specimens. Future EMCD might even have some fundamental advantages compared to

XMCD. On the one hand, the spatial resolution of a TEM outperforms XMCD by at least one magnitude. And on the other hand, the availability of EMCD is better than of XMCD as no synchrotron light source is required for the measurements.

As a final point of this summary, the experimental situation in Regensburg will be discussed. At the beginning of the ChiralTEM-project in 2004, it was not foreseeable that the Regensburg transmission electron microscope would be equipped with a post-column energy filter within the timescale of the ChiralTEM project. But an energy filter was installed in the end of 2006. By this happenstance it was even possible to comprehend EMCD measurements in Regensburg. A disadvantageous availability of the TEM system due to several technical malfunctions and a lack of technical support by the manufacturer in 2007 and 2008 prevented the challenge of further experimental tasks, such as successful EMCD measurements on rare-earth elements or first experiments of EMCD in Lorentz mode.

But instead of evaluating these offline-periods too negatively, EMCD should rather be seen as a unique chance for the transmission electron microscopy in Regensburg. Regensburg was lucky to be member of a project to explore a completely new metrology tool for the first time throughout the world. Today, the combination of high-resolution transmission electron microscopy (HRTEM), energy filtered transmission electron microscopy (EFTEM), cryo electron microscopy and Lorentz microscopy (including electron holography and differential phase contrast) in one single transmission electron microscope is once in a blue moon, perhaps even unique. So Regensburg has the best chances to compete the subject EMCD.
I would be pleased if the electron microscopy in Regensburg would continue where the ChiralTEM project ceased. The experiments on a different experimental setup for EMCD are a promising start in this direction.

Bibliography

[Ake03] P. van Aken, S. Lauterbach, *Strong magnetic linear dichroism in Fe L_{32} and O K electron energy-loss near-edge spectra of antiferromagnetic hematite $\alpha - Fe_2O_3$*, Physics and Chemistry of Minerals **30**, 469, (2003)

[Ash01] N. W. Ashcroft, N. D. Mermin, *Festkoerperphysik*, Oldenbourg Wissenschaftsverlag, München, (2001)

[Aum05] A. Aumer, *Transmissionselektronenmikroskopische Untersuchungen an ferromagnetischen Materialien für das Chiral-TEM Projekt*, diploma thesis, University of Regensburg, (2006)

[Bec02] B. Becker, *Entwicklung bildgebender Sensoren zur Darstellung raeumlicher Magnetfeldverteilungen*, diploma thesis, University of Osnarbrueck, (2002)

[Bin06] M. Binder, *Magnetization dynamics of rare-earth doped magnetic films*, dissertation, University of Regensburg, Logos Verlag, Berlin, (2006)

[Bob75] A. H. Bobeck, E. Della Torre, *Magnetic Bubbles*, North Holland Publishing Co., Amsterdam, (1975)

[Bro24] L. V. De Broglie, *Recherches sur la Theorie des Quanta*, dissertation, Sorbonne, (1924)

[Bro62] W. F. Brown, *Magnetostatic Principles in Ferromagnetism*, North Holland Publishing Company, (1962)

[Bru02] W. Brunner, *Beugungsexperimente und Paarverteilungsfunktionen zur Charakterisierung der Struktur von Seltenerdmetall/Eisen-Viellagenschichten im TEM*, dissertation, University of Regensburg, Logos Verlag, Berlin, (2002)

[Car93] P. Carra, et al. *X-ray circular dichroism and local magnetic fields*, Phys. Rev. Lett. **70**, 694, (1993)

[Chi08] http://www.chiraltem.physics.at

[Cho69] K. L. Chopra, *Thin Film Phenomena*, McGraw-Hill Book Company, New York, (1969)

[Doe06] H. Doetsch, *oral communication*, University of Osnarbrueck, , (2006)

[Dun95] C. Dunnam, *Active feedback system for suppression of alternating magnetic fields*, United States Patent 5465012, (1995)

[Ege96] R. F. Egerton, *Electron Energy-Loss Spectroscopy in the Electron Microscope*, Plenum Press, New York, (1996)

[Ers75] J. L. Erskine, E. A. Stern, *Calculation of the $M_{2,3}$ magneto-optical absorption spectrum of ferromagnetic nickel*, Phys.Rev.B **12**, 5016-5024, (1975)

[For05] P. Formanek, B. Einenkel, H. Lichte, *An improved construction of electron biprism holder for the C2-aperture*, Proceedings of Microscopy Conference 2005, Davos (Switzerland), 29.8. - 2.9.2005, p 030, (2005)

[Fre87] H. Frey, G. Kienel, *Duennschichttechnologie*, VDI Verlag, Duesseldorf, (1987)

[Ful01] B. Fultz, J. M. Howe, *Transmission Electron Microscopy and Diffractometry of Materials*, Springer Verlag, Berlin, (2001)

[Gat92] Gatan: *Instruction Manual, Models 613/636 Single Tilt/Double Tilt LN2 Cooled Specimen Holders*, Gatan Inc., (1992)

[Gat08] Gatan: *http : //www.gatan.com*, (2008)

[Gei03] J. Geissler, *Magnetische Streuung an Grenz- und Viellagenschichten*, dissertation, University of Wuerzburg, (2003)

[Goe01] E. Goering, J. Will, J. Geissler, M. Justen, F. Weigand, G. Schuetz, G., *X-ray magnetic circular dichroism - a universal tool for magnetic investigations*, Journal of Alloys and Compounds 328, 14-19, (2001)

[Has08] A. Hasenkopf, *Entwicklung von Komponenten fuer einen Chiral - STEM Betrieb*, diploma thesis, University of Regensburg, (2008)

[Hau05] T. Haug, *Simultaneous Transport Measurements and Highly Resolved Domain Observation of Ferromagnetic Nanostructures*, dissertation, University of Regensburg, Logos Verlag, Berlin, (2005)

[Hea70] O. S. Heavens, *Thin Film Physics*, Methuen & Co, London, (1970)

[Heb08] C. Hebert, P. Schattschneider, S. Rubino, P. Novak, J. Rusz, M. Stoeger-Pollach, *Magnetic circular dichroism in electron energy loss spectrometry*, Ultramicroscopy 108, pp. 277-284, (2008)

[Her06] F. Herzog, *Temperaturabhaengige TEM Untersuchungen zur Magnetisierung des Seltenerdmetalls Gadolinium*, diploma thesis, University of Regensburg (2006)

[Heu05] M. Heumann, *Elektronenholografie an magnetischen Nanostrukturen*, dissertation, University of Regensburg, Logos Verlag, Berlin, (2005)

[Hil43] J. Hillier, *On microanalysis by electrons*, Physical Review **64**, 318, (1943)

[Hil44] J. Hillier, R. F. Baker, *Microanalysis by means of electrons*, Journal of Applied Physics **15**, 663, (1944)

[Hin02] D. Hintzke, *Computersimulationen zur Dynamik magnetischer Nanostrukturen*, dissertation, Gerhard-Mercator-Universitaet Duisburg, (2002)

[Hoe04] R. Hoellinger, *Statische und Dynamische Eigenschaften von ferromagnetischen Nano-Teilchen*, dissertation, University of Regensburg, (2004)

[Hol70] L. Holland, *Vacuum Deposition of Thin Films*, Chapman and Hall, London, (1970)

[Hop05] H. Hopster, H. P. Oepen, *Magnetic Microscopy of Nanostructures*, Springer Verlag, Berlin, (2005)

[Hou07] F. Houdellier, B. Warot-Fonrose, M.J. Hytch, E. Snoeck, L. Calmels, V. Serin, P. Schattschneider *New Electron Energy Loss Magnetic Chiral Dichroism (EMCD) configuration using an aberration-corrected transmission electron microscope*, Microscopy and Microanalysis 13 Suppl. 3, pp 48-49, (2007)

[Hub98] A. Hubert, R. Schaefer, *Magnetic Domains*, Springer Verlag, Berlin, (1998)

[Huf03] S. Huefner, *Photoelectron Spectroscopy: Principles and Applications*, Springer Verlag, Berlin, (2003)

[Hus06] E. Hussnaetter, *Präparation und elektronenmikroskopische Charakterisierung von ferromagnetischen Proben für das ChiralTEM-Projekt*, diploma thesis, University of Regensburg (2006)

[Jou86] H. Jouve, *Magnetic Bubbles*, Academic Press, London, (1986)

[Kai00] S. Kaiser *TEM-Untersuchungen von heteroepitaktischen Gruppe III-Nitriden*, dissertation, University of Regensburg, (2000)

[Kit02] C. Kittel, *Einfuehrung in die Festkoerperphysik*, Oldenbourg Wissenschaftsverlag, Muenchen, (2002)

[Koe02] M. Koehler, *Die magnetische Mikrostruktur atomar geschichteter Fe/Au(001)-Viellagensysteme*, dissertation, University of Regensburg, (2002)

[Koe08] J. Koehler, diploma thesis, University of Regensburg, in progress

[Lan35] L.D. Landau, E. Lifshitz, Phys. Z. Sowjetunion 8, 153-169, (1935)

[Lan01a] R. M. Langford, A. K. Petford-Long, *Preparation of transmission electron microscopy cross-section specimens using focused ion beam milling*, J. Vac. Sci. Technol. A 19(5),(2001)

[Lan01b] R. M. Langford, Y. Z. Huang, S. Lozano-Perez, J. M. Titchmarsh, A. K. Petford-Long, *Preparation of site specific transmission electron microscopy plan-view specimens using a focused ion beam system*, J. Vac. Sci. Technol. B 19(3), (2001)

[Lav08] Property of Dr. R. Lavinsky, *www.irocs.com*, permission for publication is on hand, (2008)

[Lim06] T. Limmer, *In situ TEM-Messungen zur Schaltfeldstärke gepinnter Vortices*, diploma thesis, University of Regensburg, (2006)

[LLG08] M. R. Scheinfein, *LLG Micromagnetics Simulator*, http://llgmicro.home.mindspring.com, (2008)

[Mat96] T. Matsumoto, A. Tonomura, *The phase constancy of electron waves traveling through Boersch's electrostatic phase plate*, Ultramicroscopy, 63:5-10, (1996)

[Mat08] *http://www.mathworks.de*, (2008)

[Mei06] R. Meier, *oral communication*, University of Regensburg, (2006)

[Moe56] G. Moellenstedt, H. Dueker, *Beobachtungen und Messungen an Biprisma-Interferenzen mit Elektronenwellen*, Zeitschr. f. Physik Bd. 145, 377-397, (1956)

[Nel99] M. Nelhiebel, P. H. Louf, P. Schattschneider, P. Blaha, K. Schwarz, B. Jouffrey, *Theory of orientation-sensitive near-edge fine-structure core-level spectroscopy*, Phys. Rev. B, 59(20):12807-12814, (1999)

[Nie93] H. Niedrig, *Bergmann-Schaefer, Lehrbuch der Experimentalphysik - Optik*, Walter de Gruyter Verlag, Berlin, (1993)

[Ott01] S. Otto, *Konstruktion, Bau und Charakterisierung eines Magnet-Proben-Halters für das Elektronen-Mikroskop*, diploma thesis, University of Regensburg, (2001)

[Per04] K. Perzlmaier, *Spin Dynmics in Permalloy Microstructures in a Closure Domain State*, diploma thesis, University of Regensburg, (2004)

[Pla92] H. Plank, *Diffusionsuntersuchungen an Tb/Fe Mehrlagenschichten mit Hilfe der Auger-Elektronenspektroskopie*, dissertation, University of Regensburg, (1992)

[Rei97] L. Reimer, *Transmission Electron Microscopy*, Springer Verlag, Berlin, (1997)

[Rub07a] S. Rubino, *Magnetic Chiral Dichroism in the Transmission Electron Microscope*, dissertation, Vienna University of Technology, (2007)

[Rub07b] S. Rubino, P. Schattschneider, M. Stoeger-Pollach, C. Hebert, J. Rusz, L. Calmes, B. Warot-Fonrose, F. Houdellier, V. Serin, P. Novak, *EMCD: Magnetic Chiral Dichroism in the Electron Microscope*, Journal of Materials Research, 1026E, C13-05, (2007)

[Rub07c] Rubino, S., *oral communication*, TU Vienna, (2007)

[Rus07] J. Rusz, S. Rubino, P. Schattschneider, *First-principles theory of chiral dichroism in electron microscopy applied to 3d ferromagnets*, Physical Review B 75, 214425, (2007)

[Rus08a] J. Rusz, P. Novak, S. Rubino, C. Hebert, P. Schattschneider, *Magnetic Circular Dichroism in Electron Microscopy*, Acta Physica Polonia A 113, 599, (2008)

[Rus08b] Jan Rusz, oral communication, (2008)

[Scha03] C. Hebert, P. Schattschneider, *A proposal for dichroic experiments in the electron microscope*, Ultramicroscopy 96, 463-468, (2003)

[Scha06] P. Schattschneider, S. Rubino, C. Hebert, J. Rusz, J. Kunes, P. Novak, E. Carlino, M. Fabrizioli, G. Panaccione, G. Rossi, *Detection of magnetic circular dichroism using a transmission electron microscope*, Nature 441, 486-488, (2006)

[Scha08] P. Schattschneider, C. Hebert, S. Rubino, M. Stoeger-Pollach, J. Rusz, P. Novak, *Magnetic circular dichroism in EELS: Towards 10 nm resolution*, Ultramicroscopy, 108, 5; pp. 433 - 438, (2008)

[Schi81] G. Schimmel, W. Vogell, *Methodensammlung der Elektronenmikroskopie*, Wissenschaftliche Verlagsgesellschaft Stuttgart, (1981)

[Schn02] M. Schneider, H. Hoffmann, S. Otto, T. Haug, J. Zweck, *Stability of magnetic vortices in flat submicron permalloy cylinders*, J. Appl. Phys. 92, 1466-1472, (2002)

[Scho02] A. Scholl, H. Ohldag, F. Nolting, J. Stoehr, H. A. Padmore, *X-ray photoemission electron microscopy, a tool for the investigation of complex magnetic structures*, Rev. Sci. Instrument., **73**, 1362-1366, (2002)

[Schu87] G. Schuetz et al., *Absorption of circularly polarized x rays in iron*, Phys. Rev. Lett. **58**(7):737-740, (1987)

[Spe03] J. C. H. Spence, *High-Resolution Electron Microscopy*, Oxford University Press, (2003)

[Ste05] F. Steinbauer, *Statisches Magnetisierungsverhalten von Mikro- und Nanomagneten*, dissertation, University of Regensburg, Der andere Verlag, Toenning, 2005

[Tec08] FEI, *Tecnai online help manual - Working with a FEG*

[Tho77] K. C. Thompson-Russell, J. W. Edington, *Electron Microscope Specimen Preparation Techniques in Materials Science*, Philips Technical Library, The Macmillan Press LTD., London, (1977)

[Tho92] B. T. Thole et al. *X-ray circular dichroism as a probe of orbital momentum*, Phys. Rev. Lett. **68**, 1943, (1992)

[Tie08] T. Tietze, M. Gacic, G. Schuetz, G. Jakob, S. Brueck, E. Goering, *XMCD studies on Co and Li doped ZnO magnetic semiconductors*, New J. Phys. 10, 055009, (2008)

[Tip94] P. A. Tipler, *Physik*, Spektrum Akademischer Verlag, Heidelberg, (1994)

[Twe08] Twente Micro Products, Enschede, *http://www.microproducts.nl*

[Uhl04] T. Uhlig, *Differentielle Phasenkontrastmikroskopie an magnetischen Ringstrukturen*, dissertation, University of Regensburg, Logos Verlag, Berlin, (2004)

[Voe99] E. Voelkl, L. F. Allard, D. C. Joy, *Introduction to Electron Holography*, plenum publishers, New York, (1999)

[Wie2k] *http://www.wien2k.at*, (2008)

[Wil96] D. B. Williams, C. B. Carter, *Transmission Electron Microscopy IV - Spectrometry*, Plenum Press, New York, (1996)

[Wol04] G. Woltersdorf, *Spin-Pumping and two magnon scattering in magnetic multilayers*, dissertation, Simon Fraser University, (2004)

[Yua97] J. Yuan, N. K. Menon, *Magnetic linear dichroism in electron energy loss spectroscopy*, Journal of Applied Physics **81**, 5087, (1997)

[Zim95] T. Zimmermann, *Untersuchung der magnetischen Struktur von Co/Cu Mehrlagenschichten mit lorentzmikroskopischen Methoden*, dissertation, University of Regensburg, (1995)

A. List of Publications

A.1. Review journal paper

- P. Schattschneider, M. Stoeger-Pollach, S. Rubino, M. Sperl, C. Hurm, J. Zweck, J. Rusz, *Detection of magnetic circular dichroism on the two-nanometer scale*, Physical Review B, 78, 104413, 2008

A.2. Posters

- C. Hurm, C. Dietrich, K. Perzlmaier, A. Aumer, F. Schiller, J. Zweck, *Preparation and Characterization of magnetic specimens for ChiralTEM experiments*, Proceedings of Microscopy Conference 2005, ISSN 1019-6447, p366, 6. Dreiländertagung, Davos (Suisse), August 29 - September 2, 2005

- C. Hurm, J. Zweck, *Objective lens current reversal for ChiralTEM measurements*, Proceedings of the 16th International Microscopy Congress, P8M 169, Sapporo (Japan),September 3-8, 2006

A.3. Talks at international meetings

- C. Hurm, M. Heumann, T. Uhlig, J. Zweck (INVITED), *Experimental determination of the magnetic detection limit of Electron Holography in comparison with DPC*, Proceedings of Microscopy Conference 2005, ISSN 1019-6447, p34, 6. Dreilaendertagung, Davos (Suisse), August 29 - September 2, 2005

- C. Hurm, M. Stoeger-Pollach, S. Rubino, C. Hebert, P. Schattschneider, J. Zweck, *Verification of Electron Magnetic Chiral Dichroism in a TEM by Reversing the Specimen's Magnetisation, Microscopy and Microanalysis*, Vol. 13, Supp. 3, Saarbruecken, September 2-7, 2007

A.4. Talks at ChiralTEM meetings and workshops

- C. Hurm, J. Zweck, *ChiralTEM interim report, University of Regensburg*, ChiralTEM meeting, Prague (Czech Republic), April 8, 2005

- C. Hurm, C. Dietrich, K. Perzlmaier, A. Aumer, F. Schiller, J. Zweck, *Preparation and Characterization of magnetic specimens for ChiralTEM experiments*, 1st ChiralTEM workshop, Davos (Suisse), August 29, 2005

- C. Hurm, J. Zweck, *Specimen preparation and control over magnetic fields for ChiralTEM experiments*, 2nd ChiralTEM workshop, Vienna (Austria), April 19, 2006

- C. Hurm, E. Hussnaetter, J. Zweck, *Specimen preparation and control over magnetic fields for ChiralTEM experiments*, ChiralTEM meeting, Dresden, October 27, 2006

- C. Hurm, A. Schneider, J. Zweck, *Control over magnetic fields for ChiralTEM experiments*, 3rd ChiralTEM workshop, Trieste (Italy), May 31, 2007

B. Pulsed field specimen holder

This appendix contains construction data of the pulsed field specimen holder discussed in section 8.3.

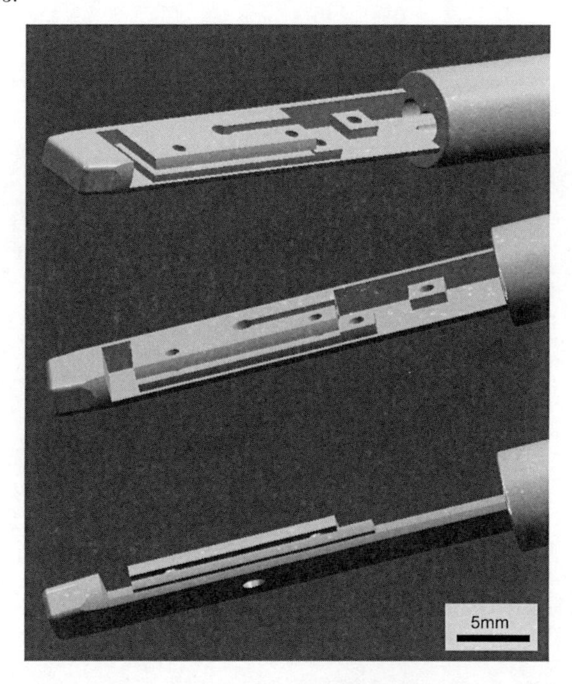

Fig. B.1.: *Tip of the pulsed field specimen holder. This illustration may help to comprehend the mechanical drawings on the following pages.*

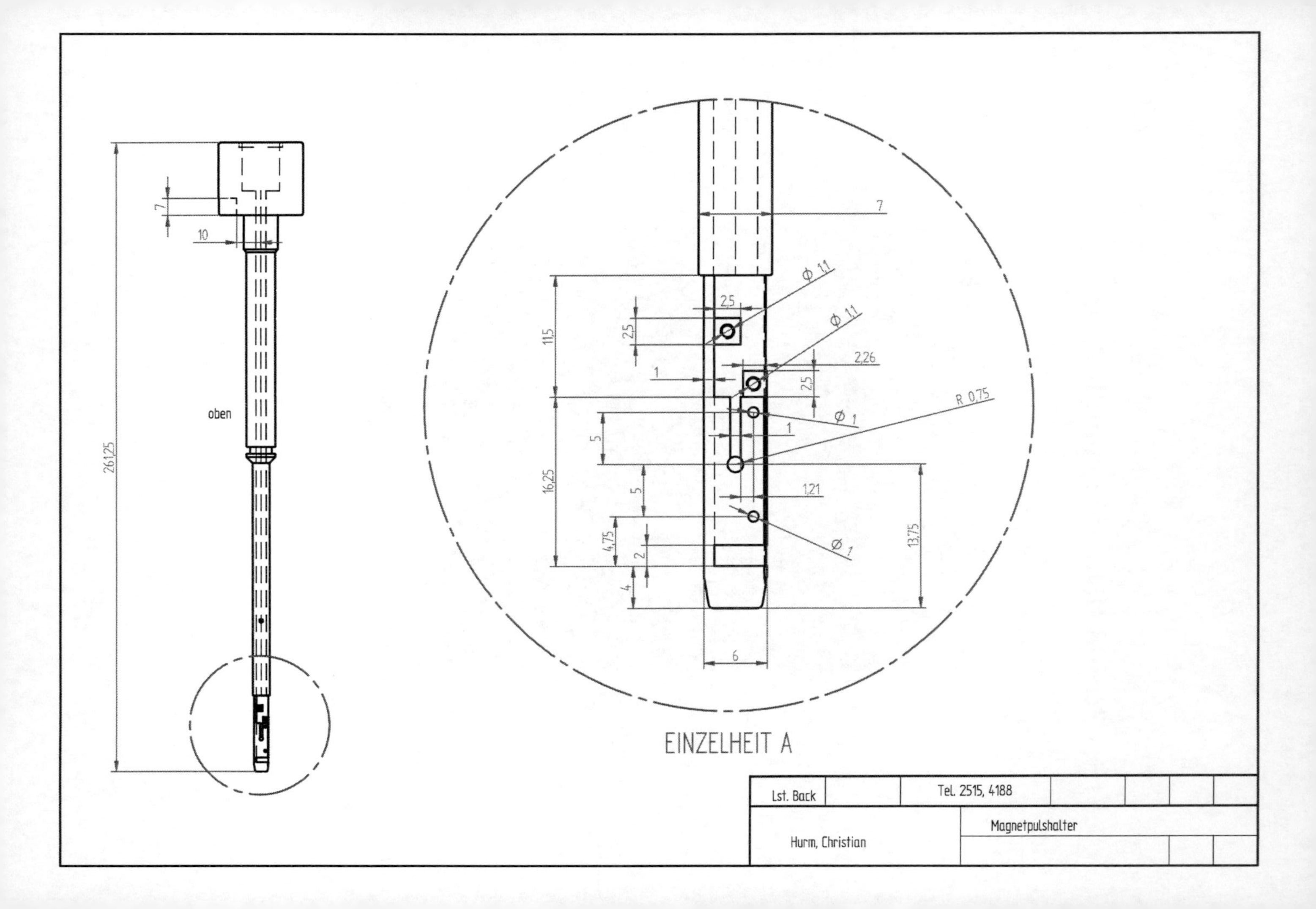
261,25
7
10
oben
7
Φ 1,1
2,5
Φ 1,1
11,5
2,5
2,26
1
2,5
R 0,75
Φ 1
5
1
16,25
5
1,21
13,75
4,75
2
Φ 1
4
6
EINZELHEIT A
Lst. Back
Tel. 2515, 4188
Magnetpulshalter
Hurm, Christian

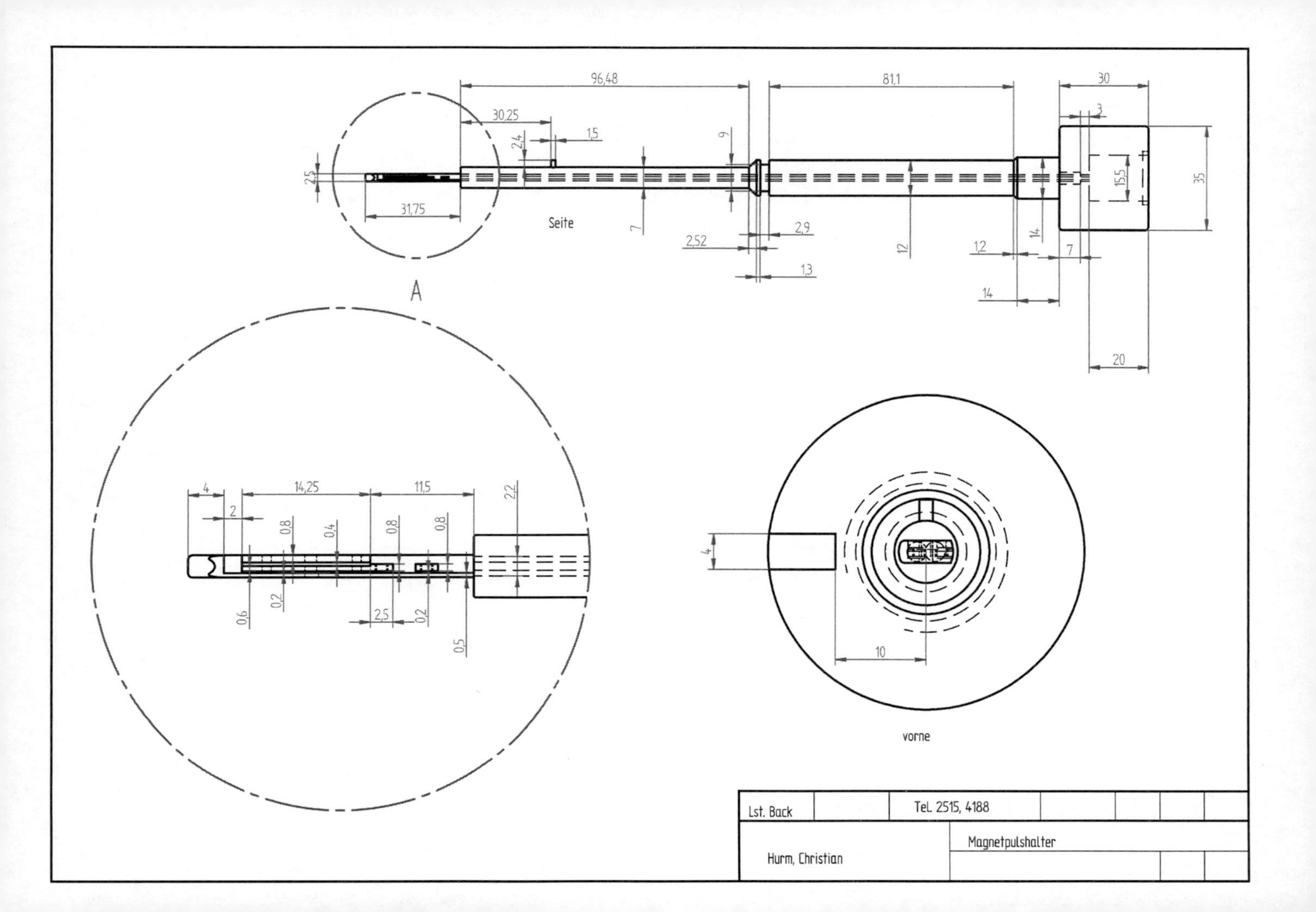
96,48
81,1
30
30,25
3
2,4
1,5
9
2,5
15,5
35
31,75
Seite
7
2,52
2,9
12
1,2
14
7
1,3
14
20
A
4
14,25
11,5
2,2
2
0,8
0,4
0,8
0,8
0,6
0,2
2,5
0,2
0,5
4
10
vorne
Lst. Back
Tel. 2515, 4188
Magnetpulshalter
Hurm, Christian

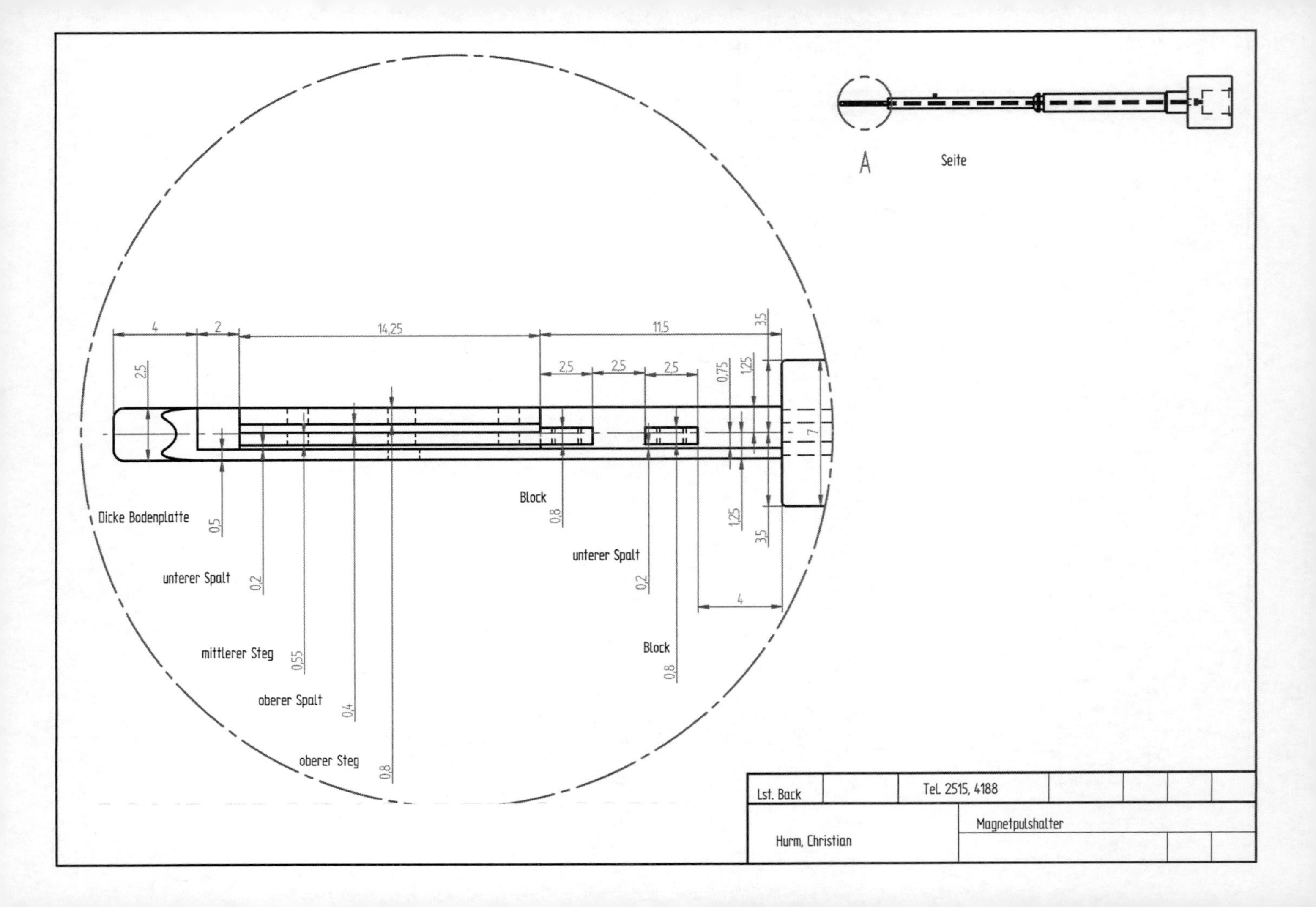
A
Seite
4
2
14,25
11,5
3,5
2,5
2,5
2,5
2,5
0,75
1,25
7
Block
0,8
1,25
3,5
Dicke Bodenplatte
0,5
unterer Spalt
0,2
unterer Spalt
0,2
4
mittlerer Steg
0,55
Block
0,8
oberer Spalt
0,4
oberer Steg
0,8
Lst. Back
Tel. 2515, 4188
Magnetpulshalter
Hurm, Christian

C. Commutation Unit

This appendix contains construction data of the commutation unit discussed in section 7.2.1. The figures on this page show peripherals, the following two pages show a circuit diagram of the latest version of the commutation unit.

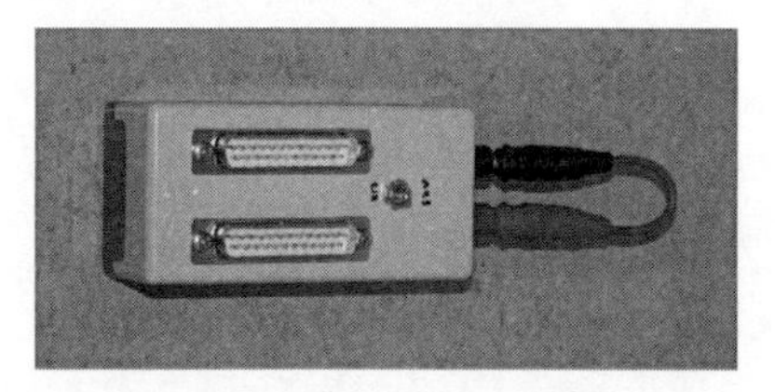

Fig. C.1.: *The* ***lens control unit*** *can interrupt all lens currents, either manually by changing over the switch or automatically by a control signal of the commutation unit (the connections to the commutation unit are bridged in this image).*

Fig. C.2.: *The* ***bridgeover unit*** *is a passive jumper that can replace the commutation unit quickly if it averts suspicion to cause problems.*

Uni Regensburg Elektronik - Physik	ep517-3Polwender	
	Auftrag 06112	
	EP - Nummer:	Bearbeiter: Alexander Meier
Datum: Nov.06	Datei: Mainboard0307Do.sch	Karl Dorsch

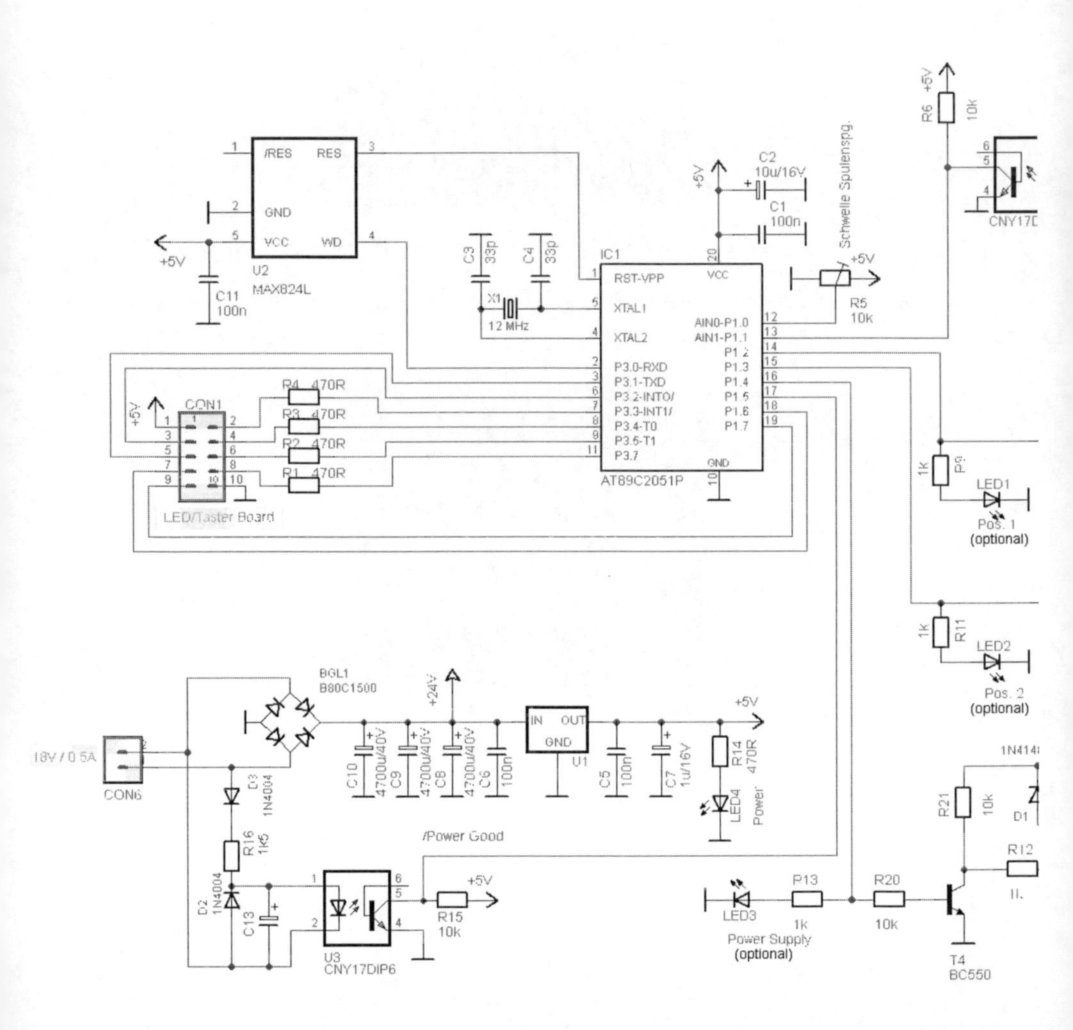

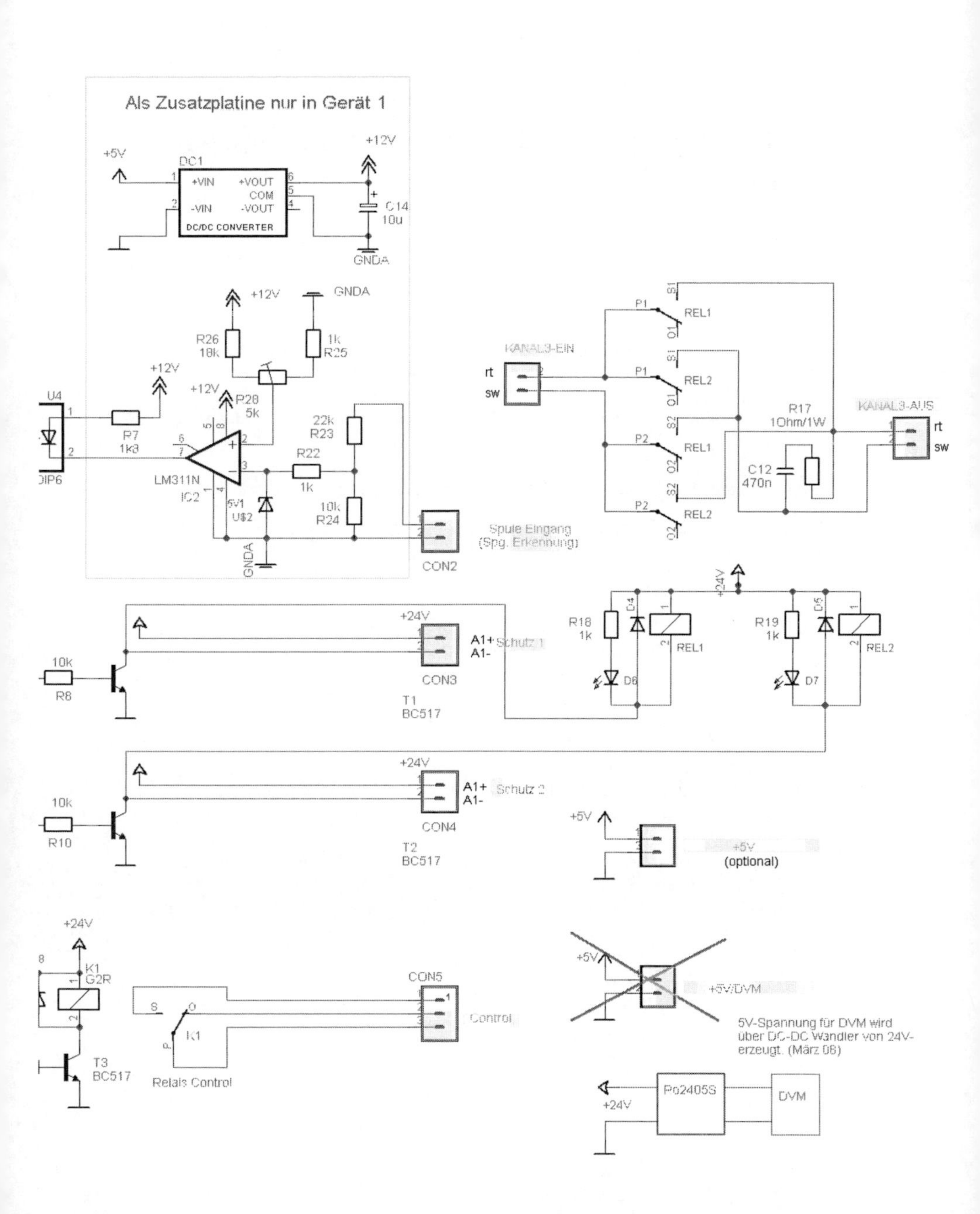

Als Zusatzplatine nur in Gerät 1
DC/DC CONVERTER
Spule Eingang
(Spg. Erkennung)
KANAL3-EIN
KANAL3-AUS
Schutz 1
Schutz 2
+5V
(optional)
+5V/DVM
5V-Spannung für DVM wird
über DC-DC Wandler von 24V-
erzeugt. (März 08)
Po2405S
DVM
Control
Relais Control

D. GIF alignment for the spectrum spread method

Guide for the spread of the electron energy loss pattern[1] on the energy filter CCD (Gatan imaging filter) in the direction perpendicular to the energy loss axis in key points:

→ Be sure to have the GIF settings saved before making changes in the service menu!

→ Start *Tecnai User Interface*, *Gatan Filter Control* and *Gatan Digital Micrograph* (in this order)

→ Chose *calibrate* → *service mode* in the filter control window

→ After the acoustic signal, type "gatan" (without prompt or input window!)

→ Chose *Windows* → *Post Slit* to open the post slit lens setup window

→ double-klick on *Disperion* to select or create an energy dispersion value for the intended changes. Ideally chose a nomenclature that differs apparent from the normal setup, for example by chosing values as 0.11 eV instead of 0.10 eV

→ The single lens currents can be changes after double-clicking the particular lens name

→ The electron pattern on the energy filter CCD is spread in y-direction by setting the value of the Q6 lens to zero.

→ The spectrum is shifted in the energy loss direction (x) by changing the value of the D3X lens. Ideally the center of the zero loss peak should be around pixel 100 (in x direction, beginning from the left side) - To receive conventional spectra with modified lenses, the readout area has to be adjusted changing *Gatan Digital Micrograph* → *Filter* → *Spectroscopy Setup* → *Readout Options.*

[1]This modification simplifies measurements using the spectrum spread method shown in section 3.3.2

E. Acknowledgment

Ein herzliches Dankschön an alle, die mich bei dieser Arbeit unterstützt haben - und ganz besonders an...

... Joe Zweck für die Themenstellung und die vertrauensvolle Zusammenarbeit.

... Jürgen Gründmayer und Christian Dietrich für die stets angenehme Atmosphäre im Büro und sämtliche kleinen und großen Hilfestellungen.

... die wechselnden Kollegen am vierten Schreibtisch im Büro, Stephanie Hußnätter, Christoph Kefes, Roman Grothausmann und Martin Wengbauer.

... die "ChiralTEM Diplomanden" Eva Hußnätter und Andreas Hasenkopf.

... alle anderen aktiven und ehemaligen Mitglieder der TEM-Gruppe, Martin Beer, Michael Binder, Martin Brunner, Karl Engl, Thomas Haug, Florian Herzog, Martin Heumann, Michael & Christian Huber, Sebastian Liebl, Marcello Soda, Johannes Thalmair, Thomas Uhlig, und ganz besonders an den "guten Geist" im Labor, Olga Ganitcheva.

... alle Beteiligten der ChiralTEM Partnergruppen in Wien, Prag, Dresden und Triest.

... Korbinian Perzlmaier für diverse Hilfestellungen, insb. bei den Simulationen.

... Björn Becker und die Gruppe von Prof. Doetsch für die YIG-Proben.

... Dr. Roland Kröger für die FIB-Schnitte.

... Roland Meier für die GaAs-Proben.

... meine Eltern, die mich während der gesamten Zeit vertrauensvoll unterstützt haben.

Ganz besonders möchte ich mich bei meiner lieben Tina bedanken, die mich während der langen Zeit des Zusammenschreibens geduldig ertragen und immer wieder neu motiviert hat.

Bisher erschienene Bände der Reihe
„Applied Electron Microscopy - Angewandte Elektronenmikroskopie“

ISSN 1860-0034

Nr.	Autor	Titel	ISBN	Preis
1	Wolfgang Brunner	Beugungsexperimente und Paarverteilungsfunktionen zur Charakterisierung der Struktur von Seltenerdmetall/Eisen-Viellagenschichten im TEM	ISBN 3-89722-895-5	40.50 EUR
2	Markus Schneider	Untersuchung des Ummagnetisierungsverhaltens von polykristallinen Mikro- und Nanostrukturen aus Permalloy mit der Lorentz-Transmissionselektronenmikroskopie	ISBN 3-8325-0187-8	40.50 EUR
3	Thomas Uhlig	Differentielle Phasenkontrastmikroskopie an magnetischen Ringstrukturen	ISBN 3-8325-0752-3	40.50 EUR
4	Martin Heumann	Elektronenholografie an magnetischen Nanostrukturen	ISBN 3-8325-0870-8	40.50 EUR
5	Karl Engl	TEM-Analysen an Gruppe III-Nitrid Heterostrukturen	ISBN 3-8325-1102-4	40.50 EUR
6	Thomas Haug	Simultaneous Transport Measurements and Highly Resolved Domain Observation of Ferromagnetic Nanostructures	ISBN 978-3-8325-1328-3	40.50 EUR
7	Christian Hurm	Towards an unambiguous Electron Magnetic Chiral Dichroism (EMCD) measurement in a Transmission Electron Microscope (TEM)	ISBN 978-3-8325-2108-0	40.50 EUR

Alle erschienenen Bücher können unter der angegebenen ISBN direkt online (http://www.logos-verlag.de/Buchreihen) oder per Fax (030 - 42 85 10 92) beim Logos Verlag Berlin bestellt werden.